RADIOLOGIC
QUALITY CONTROL MANUAL

RADIOLOGIC QUALITY CONTROL MANUAL

Triphy C. Barber, R.N., B.S., M.S., Ph.D

Mike Thomas, R.T., B.S., R.D.M.S.

RESTON PUBLISHING COMPANY

A Prentice-Hall Company

Reston, Virginia

Library of Congress Cataloging in Publication Data

Barber, Triphy C.
Radiologic quality control manual.

Bibliography: p
1. Radiography, Medical—Quality control—Handbooks, manuals, etc. I. Thomas, Mike. II. Title. [DNLM: 1. Radiology—Standards—Handbooks. 2. Quality control—Handbooks. WN 21 B234r]
RC78.B325 1983 616.07'52'0289 83-2971
ISBN 0-8359-6366-7

Editorial/production supervision and interior design by Camelia Townsend

10 9 8 7 6 5 4 3 2 1

PRINTED IN THE UNITED STATES OF AMERICA

This manual is dedicated to:

All the Radiologists and Radiology Technologists whom I have had the privilege of working with.

T.B.

My wife and my parents, who have given me encouragement and support through the years.

M.T.

CONTENTS

PREFACE

It should be the goal of every radiology facility to have the best possible radiographs at the lowest possible radiation dose levels for the patient and staff. Quality control and, in a larger sense, the quality assurance program provide a means of obtaining this goal. Quality control in the Diagnostic Radiology Department is a many faceted program involving each member of the radiographic team. This manual is written to aid in the establishment and/or upgrading of radiographic quality control. Furthermore, it is the goal of the authors, in writing this manual, to assist radiologic departments to meet their goal of optimum radiographs at optimum safety for the patient and staff.

ACKNOWLEDGMENTS

We wish to thank the following people and companies for their assistance and cooperation in completing this manual: Clifford W. Barber; Randii Branch, Dot Langfitt; E. I. duPont de Nemours and Co., Inc.; Victoreen Nuclear Associates; Shielding, Inc.; Dr. Gopala Rao; Supertech, Inc.; and Ilford, Inc.

INTRODUCTION

Purpose of the Manual

It is the purpose of a quality control manual to familiarize everyone with the concepts and functions of radiographic quality control and the radiographic quality assurance program.

1. It is each quality control team member's responsibility to know what his or her duties are and to perform them accurately.
2. It is each team member's responsibility to know what each of the quality control procedures are in the Radiology Department and to understand the principles and methodology of the procedures.
3. The *team* approach is essential to a well functioning quality control and quality assurance program. Learn as many things as you can concerning quality control and the inner workings of your area. The more proficient you become, the more valuable person you become to the patient, department, and hospital.
4. It is very important that everyone in the quality control and assurance program function as a team. Your team will work only as well as each part of the team works.
5. It is of the utmost importance that you not be absent from work except in the most extreme emergency. Please call in early if you are unable to come to work.

Need for the Manual

The Quality Control Manual serves several purposes:

1. It lists quality control policies and procedures.
2. It provides information on radiographic quality control and quality assurance programs.
3. It serves as a training aid for new technologists.
4. It delineates responsibility of team members.
5. It containes a list of all the parameters to be monitored in the diagnostic radiographic facility. This includes the frequency at which each parameter is to be monitored and by whom it is to be monitored.
6. A detailed procedure for each radiographic parameter to be monitored is included in the manual.

7. Steps to troubleshoot problems encountered in quality control are identified.
8. A list of records to be kept for quality control and sample forms are in the manual.
9. The development of criteria and standards by which new equipment is to be evaluated is outlined.
10. References for further assistance and guidelines are listed.

RADIOGRAPHIC QUALITY CONTROL POLICIES

1. ______________________ hospital will institute Radiographic Quality Control and a Quality Assurance Program on ______________________.
2. Those responsible for the quality control and quality assurance program are:
   a. *Physician Director of Radiology:* As head of the department the physician director is ultimately responsible for all programs in the department.
   b. *Chief Technologist:* The responsibility for coordinating quality control programs in the Department of Radiology is usually delegated to the chief technologist.
   c. *Quality Control Coordinator:* The actual responsibility of doing daily quality control is designated to an individual in the X-ray Department. This person is usually given the title of the quality control coordinator.
   d. *Radiographic In-service Educator:* It is this person's responsibility to conduct a current in-service program in quality control.
   e. *In-house and/or Contract Service:* It is the responsibility of the service department to keep the equipment in excellent condition through preventive and corrective maintenance.
   f. *Physicist:* The physicist is responsible for checking the x-ray system to determine that it meets all safety standards for the patient, physician, and technologists. Many hospitals that do not have a physicist on service will share a physicist with other hospitals in the area.
3. Individuals listed above constitute the quality assurance committee. It shall meet a minimum of four times a year, with additional meetings as needed.
4. The quality control coordinator is responsible for quality control of each radiographic process as outlined by the quality control committee. Radiographic processors shall be checked at the start of each day and the results documented.
5. Items to be checked during processor quality control:
   a. Sensitometric film strips.
   b. Film lot number.
   c. Replenishment rates of the processors.
   d. Temperature of the developer, water, and dryer.
   e. Densitometer data, with documentation of data points on control charts.
6. Frequency checks for quality control of other radiographic equipment are established by the quality assurance committee.

7. Quality control and the quality assurance programs are reviewed four times a year. Additional meetings are called as necessary for problem areas associated with quality control. The review committee includes the quality control committee and additional members from medical records and administration.

RADIOLOGIC QUALITY CONTROL IN-SERVICE EDUCATION PROGRAM

Overview

It is important that the radiologic in-service education program for quality control be approached from two directions. This methodology includes: (1) training prior to the start of the quality assurance program and, (2) continuing education to keep all personnel abreast of new developments in quality control.

Radiologic in-service education must include training *all* personnel who have quality control or assurance responsibilities. Practical experience with quality control techniques conducted either by an experienced instructor or by the "buddy" system is the most desirable method of training. If supervised instruction is not immediately available, self-instruction (programmed learning) may be adequately substituted until supervised instruction is available.

QUALITY CONTROL MAINTENANCE PROGRAM

Overview

The maintenance program is an important aspect of the radiographic quality control and quality assurance program. The maintenance program is broken down into two parts, preventive maintenance and corrective maintenance.

PREVENTIVE MAINTENANCE

The key word to preventive maintenance is *regularity*. The objective for preventive maintenance is to maintain minimal down time of a radiographic system. The goal of preventive maintenance is to prevent breakdowns due to equipment failure without warning.

Everyone can assist in preventive maintenance. Visual inspection of equipment before use, not using equipment until instructed in its proper use, following the manufacturer's instructions and recommendations for use, cleaning of the equipment, and taking care of the equipment after use promotes preventive maintenance.

When setting up a preventive maintenance schedule for an x-ray system, it is helpful to schedule a regular day and time. This allows adequate time for the service engineers to properly check the equipment. This system also allows scheduling of patients around the maintenance period.

CORRECTIVE MAINTENANCE

Corrective maintenance can be kept to a minimum with a good quality control preventive maintenance program. As soon as potential problems or actual problems are ascertained, corrective action is instituted according to the policies of the radiologic facility.

DOCUMENTATION AND RECORDS FOR QUALITY CONTROL AND QUALITY ASSURANCE PROGRAM

Overview

A method of documentation and recording monitoring techniques, problems encountered, preventive and corrective maintenance, and the effectiveness of these measures is needed. These records serve as an instrument by which the quality control program may be evaluated. The program may be used as a starting place for troubleshooting new problems in the future.

QUALITY CONTROL EVALUATION

Overview

To be effective, quality control should be evaluated by the performance of the x-ray systems and film quality. An evaluation of the quality assurance program itself is also needed.

X-RAY SYSTEM EVALUATION PARAMETERS

This evaluation includes:

1. Comparison of monitoring data with the purchase specifications of the x-ray system.
2. Analysis of trends in monitoring data.
3. Data used in initiating corrective action.
4. Acceptance testing results.

QUALITY ASSURANCE PROGRAM

This evaluation includes:

1. Analysis of technical error.
2. Amount of equipment repair and reasons for the repair.
3. Cost of new equipment.
4. Complaints of physicians, radiologists, and technologists.
5. Subjective evaluation of physical image quality.
6. Effect of quality control monitoring sensitometric strips.
7. Miscellaneous problems.

REVIEW PROCESS FOR RADIOGRAPHIC QUALITY CONTROL AND QUALITY ASSURANCE PROGRAM

Overview

Review is an important aspect of the quality control program. It is through this process that the effectiveness of the program is determined. The shortcomings of the program as well as its outstanding qualities are identified. The items that need to be reviewed are as follows:

1. Records of preventive maintenance monitoring and corrective maintenance.
2. The appropriateness of monitoring and maintenance techniques.
3. Established operating standards, both objective and subjective.
4. The evaluation of the quality control program to determine its effectiveness.
5. Review and revision of the quality control manual as necessary.

REVIEW COMMITTEE

1. Radiologist assigned to quality control
2. Quality Control Coordinator
3. Administrative Head of the Radiology Department
4. Radiology In-service Education Coordinator
5. Employee from medical records
6. Physician's Representative

FREQUENCY OF PROGRAM REVIEW

A quality control review committee should meet at a minimum of four times per year. The length of time the program has been operational, the number of problems, and the size of the facility determine the need for additional review.

RESPONSIBILITIES OF THE QUALITY CONTROL COORDINATOR

The responsibilities of the quality control coordinator are as follows:

1. To assume responsibility for the quality control equipment.
2. To initiate quality control each morning on specific pieces of equipment as outlined by the quality assurance committee.
3. To document daily quality control by the use of control charts, graphs, and the like.
4. To complete quality control checks on other pieces of radiographic equipment as determined by the quality assurance committee.
5. To take appropriate, corrective action if quality control values lie outside the established control limits.
6. To contribute as a member of the quality assurance committee.
7. To assist in the evaluation and recommendations of new products.
8. To assist in preparation and instruction of the in-service quality control program.
9. To assist in maintenance of the quality control manual, equipment, and program.
10. To take an active role in the review process for quality control and the quality assurance program.

PHYSICAL AREA FOR QUALITY CONTROL

Overview

An area for all quality control equipment and all necessary records is essential. One logical place is a darkroom (preferably one that is not extremely busy). Ultrasound or nuclear darkrooms are excellent examples of darkrooms that might be utilized for quality control. A second area that may function as a quality control center is a small office in close proximity to the radiographic processors. The quality control equipment remains set up at all times. An area to file the quality control records is also needed.

TROUBLESHOOTING CHAIN OF COMMAND

Overview

When corrective action is needed to bring quality control within the established control limits, it is wise to have an established procedure.

CHAIN OF COMMAND

1. *Quality Control Coordinator:* Recognizes the problem and initiates appropriate action.
2. *Chief Technologist:* Second in the chain of command. He or she may take the responsibility of calling in service or may troubleshoot the problem personally.
3. *Physician Department Head:* If there is an interruption in service, the department head should be informed of the problem.
4. *In-house and/or Contract Service:* Called in to troubleshoot major calibration or quality control problems.
5. *Quality Control Coordinator:* Rechecks quality control after service and documents the results.
6. *Administrative Department Head:* Becomes involved when there is a breakdown in communications or a need for administration action.

RADIOGRAPHIC QUALITY CONTROL AND THE QUALITY ASSURANCE PROGRAM

Overview

In the last 25 to 30 years, there have been significant technological advances in the photographic processing of medical radiographs. In this era, the field of radiology has seen the change from manual processing of radiographs to the invention of the rapid process automatic film processor by Eastman Kodak in the late 1960s. The need for clinical quality control as well as radiographic quality assurance programs is now being recognized.

WHAT IS RADIOGRAPHIC QUALITY CONTROL?

Radiographic quality control involves the use of specific activities and equipment to establish operating standards and to monitor radiation output from x-ray systems. It also involves the recalibration of the radiographic equipment to meet the established operating standards.

WHAT IS A RADIOGRAPHIC QUALITY ASSURANCE PROGRAM?

A radiographic quality assurance program includes quality control, an active in-service education program for all radiology personnel, a preventive and corrective maintenance program, evaluation of new products, and a documentation and review program. When all these areas, which constitute the quality assurance program, are integrated successfully, the result is the highest diagnostic quality, the lowest patient radiation dosage, and the most efficient low cost for the radiology department.

WHY IS A RADIOGRAPHIC QUALITY ASSURANCE PROGRAM NECESSARY?

Optimum diagnostic quality is provided through a working quality control and preventive maintenance program.

WILL A QUALITY ASSURANCE PROGRAM PROMOTE COST CONTAINMENT?

Yes. Radiation dose reduction, optimum diagnostic image quality, and a reduction in the number of repeated radiographs due to technical error result in a highly productive and cost efficient program for the radiology department and, in turn, the hospital.

A formula can thus be derived and documented concerning quality control cost containment:

$$\frac{\text{Minimum radiation dose}}{\text{Optimum image quality}} = \text{Minimum Amount of Dollars Spent}$$

Other areas that promote cost containment within the radiology facility are better utilization of the radiographic rooms and personnel, and the potential savings of expendable products.

There is another intangible (subjective) item that has a direct effect on cost containment and specifically the radiographic quality assurance program: the mutual trust and reliance between the x-ray technologists and the radiologists. Through the quality assurance program, this relationship can be nurtured to the benefit of the patient, technologist, radiologist, physician, radiology department, and the hospital as a whole.

DEFINITIONS APPLICABLE TO QUALITY CONTROL

Accuracy Precision.

Automatic Film Processor Mechanized method of transporting a film from developer to fixer to a water wash and to the dryer. This process results in a radiograph that is ready to read.

Base Fog Inherent blackness within the base of the film.

Base Plus Fog Inherent blackness within the base of the film and minimal chemical fog produced by processing.

Calibration To fix, check, or correct the scale of a measuring instrument.

Cassettes Light-tight box or holder for the radiographic film and screen.

Congruence The quality of being in harmony or agreement (e.g., film-screen congruence).

Contrast The differences between photographic densities (white to black).

Control Chart Graphs utilized for recording data for accurate analysis and interpretation.

Control Limits The acceptable limits for monitoring processor technique (range plus/minus 0.10 to 0.15 density and plus/minus 0.5 base plus fog).

Control Strips Film that has been exposed by a sensitometer or radiation utilizing a step wedge or phantom.

Crossover Testing The procedure by which a master control roll of film or new lot number of film is introduced into the radiographic quality control system. It involves the simultaneous running of the old film with the new film through sensitometric and densitometric readings.

Data Points The actual values obtained by using a sensitometer for exposure and measuring the optical density with a densitometer.

Densitometer An instrument designed to measure the optical density of photographic film and materials post processing and drying.

Density Degree of opacity.

Distortion To change shape, form, or appearance. The production of an unfaithful reproduction.

Drift The course on which something is directed. To move away gradually from a set point.

Emulsion A suspension of a salt of silver mixed with a gelatin or collodion to coat a film.

Established Control Standards The objective and subjective operating standards that have been personalized for your specific equipment.

Flood Replenishment Rapid exchange of chemicals over a set time interval.

Fog Any element that darkens the film that is not part of the primary radiograph.

F-Stop Equals the diameter of the lens divided by the focal length of the lens when it is focused on infinity.

Glare Smooth, bright, glassy surface.

Grid Types: crosshatched, focused, linear, parallel. Composed of strips of lead alternating with strips of aluminum or plastic spacers.

Half-Value Layer The thickness of aluminum that will cause one-half the amount of radiation to be emitted as was put into it.

Linearity Pertaining to or resembling a line or lines.

Medical Radiograph Medical photograph produced by radiation acting upon a screen resulting in light exposure of photosensitive materials.

Nonimage Radiation A source of radiation that affects the contrast of the film in addition to the primary beam.

Objective Standards Those standards concerned with actual features that may be directly measured.

Phantom A man-made instrument to simulate the physical properties of a human body. Used in the calibration of the x-ray system.

Physical Image Quality Constitutes the contrast, resolution, and noise properties of a film.

Quality Assurance The planned and systematic actions that provide adequate confidence that x-ray facilities will have the best quality, the lowest patient radiation exposure, and the lowest cost containment.

Quality Assurance Administration Management actions intended to guarantee that monitoring techniques are properly performed and evaluated, with corrective action taken as needed.

Quality Assurance Program An organized group designed to provide quality assurance for a diagnostic radiology facility.

Quality Control See *Quality Assurance.*

Quality Control Coordinator Designee who is directly in charge of quality control techniques.

Quality Control Monitoring Techniques used in testing and maintaining the components of an x-ray system.

Quantum Mottle Noise. The statistical distribution of x-ray photons across the surface of the screen, with one area receiving more photons than an adjacent area.

Radiation Measuring Devices Instruments used to measure radiation output. Examples: dosimeter and R meter.

Radiograph Lot Number The quality control number given to a batch of film by the manufacturer.

Radiology Medical Facility Any facility in which any type of x-ray system or systems are used for diagnosis or visualization of any anatomical part of the human body.

Replenishment Specific amount of chemical replacement over a specified amount of time.

Reproducibility Repeatability.

Resolution Sharpness, measured in line pairs per millimeter.

Screen Phosphorescence sheet that, when struck by radiation, emits light. There is a linear response of amount of radiation to the amount of emitted light.

Sensitometer An instrument designed to provide a reproducible, known exposure to photosensitive materials (film) through an attenuator (step wedge).

Step Wedge A graduated instrument to assist in the calibration of x-ray systems.

Subjective Standards Concerned with thoughts, feelings, and other expressed emotions that cannot be directly measured.

Trends To bend or turn in a specific direction.

User Anyone who utilizes ionizing radiation for diagnostic or therapeutic purposes.

X-ray System The reassemblage of many components for the controlled production of diagnostic images with x-ray.

RADIATION SAFETY AND EQUIPMENT

Overview

Radiologic quality control cannot be discussed without including a section on radiation safety. Radiation safety is an integral part of radiologic quality control and the quality assurance program. This section incorporates the equipment and procedures used in the practice of radiation safety.

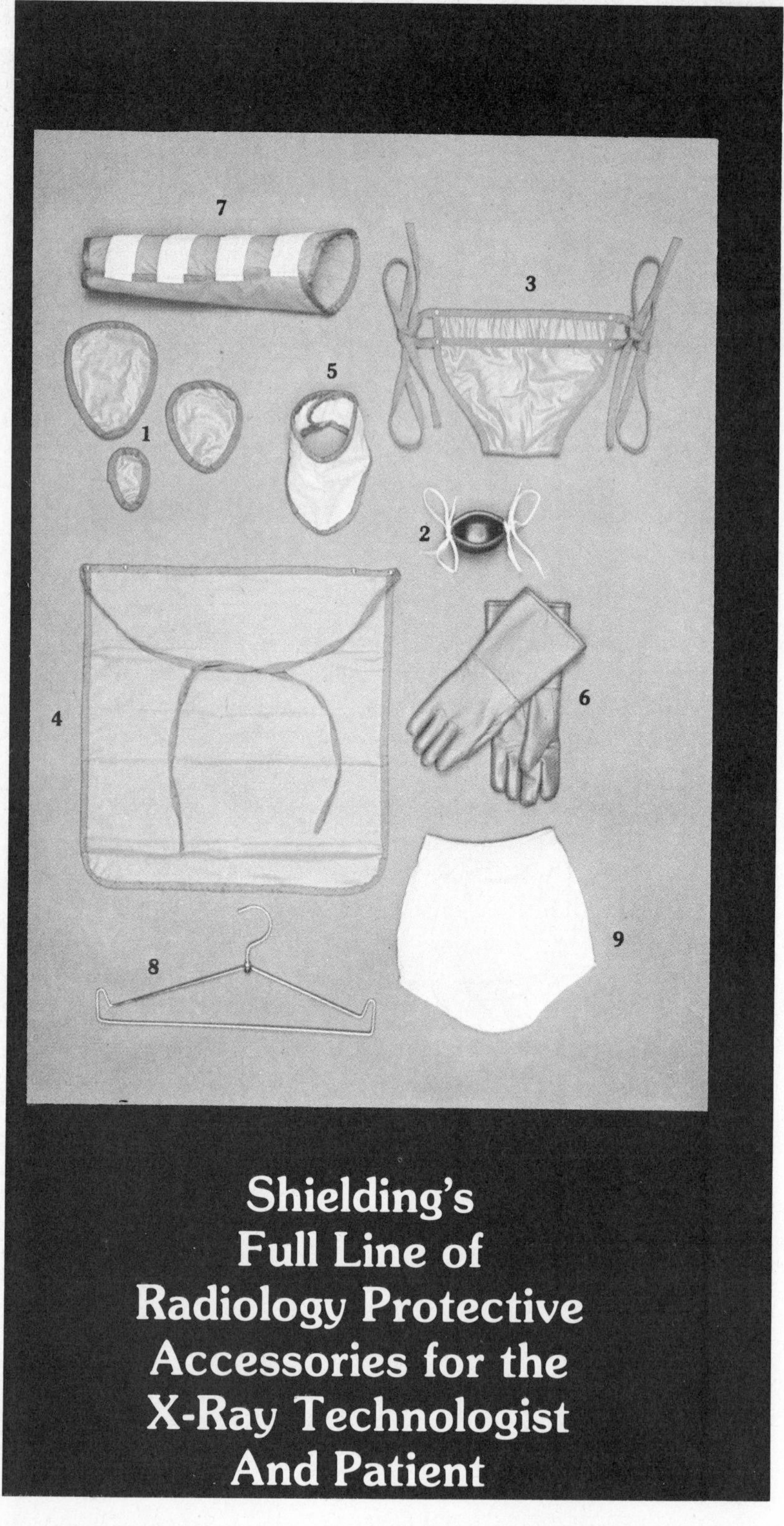

Catalog Sheet No. 6

Shielding Quality In Protective Accessories Means Extra protection for Technologists And Patients

The same kind of research and quality of production goes into Shielding's full line of radiology protective accessories that has made the firm the industry leader. The result is greater protection for technicians and patients.

For example, Shielding research has produced the industry's only one-piece lead vinyl injection molded Glove and Gonad Cup. This is one of the reasons for Shielding's continued leadership in the radiology protective garment industry.

Don't your technologists and patients deserve the full measure of Shielding quality in your radiology protective accessories?

Radiologic Protective Devices
Courtesy of Shielding, Inc., Madras, Ore.

RADIATION PROTECTIVE EQUIPMENT

Lead Aprons

- These are expensive and can crack easily if misused.
- The aprons should be hung on a rack when not in use. They **should not** be over a chair, on the floor, or folded up.
- For the protection of personnel, aprons are checked for tears and cracks every six months.

Thyroid Collars

- These can be ordered upon request for use in rooms where there is a great amount of radiation (e.g., special procedures and general fluoroscopy rooms).

Lead Glasses

- These are very expensive items. If a physician does many special procedures a month, they should be available to him.
- Purchasing of prescription glasses is the responsibility of the individual radiologist or physician.

Protective Gloves

- Protective gloves need to be worn on those procedures where the physician's hands are in the primary x-ray field.

Exposure Badges

- All technologists and physicians must wear their radiation badge on every case.
- Badges are not to be taken home.
- Badges that are lost will be replaced. The cost may be charged to the individual.

Lead Door

- This is a special shield used to cut down the amount of scatter radiation. It usually has a lead window for good visualization.

Shielding Quality
In Protective
Accessories
Means
Extra protection
For Technologists
And Patients

Compare these features . . . . compare these advantages in Shielding protective accessories for technologists and patients:

❶ GONAD SHIELDS
Full .5mm equivalency — exclusive Multi-Ply construction (a development of Shielding research). Available in three sizes. Custom sizes on order

❷ GONAD CUPS
The industry's only one-piece lead vinyl injection molded gonad cup. Full .5mm equivalency. For adult use.

❸ DIAPERS
Full .5mm equivalency. Exclusive Multi-Ply construction (a development of Shielding research). Full size — 14"x20".

❹ HALF APRONS
Full .5mm equivalency. Exclusive Multi-Ply construction (a development of Shielding research). Three sizes: 24"x24" — 20"x24" — 8"x10" (childs)

❺ LEAD COLLARS
Full .5mm equivalency. Exclusive Multi-Ply construction (a development of Shielding research). Two sizes — large and small.

❻ GLOVES
The industry's only one-piece lead vinyl injection-molded glove. Full .5mm equivalency. Three covers available — jersey sheath, fawn leather and blue leather.

❼ SLEEVE
Provides complete arm protection. Available in .3mm or .5mm equivalency. Velcro closure for fast and easy securing. Small, medium or large sizes.

❽ HEAVY-DUTY APRON HANGER
Will hold up to 50 pounds. Chrome-plated, 1/4" steel construction.

❾ PANTIES
Two pair white polyester panties plus one set of lead shields. A .5mm front shield and a .3mm back shield. Shields are removable for laundering the panties. They come in small, medium and large sizes.

LEAD CURTAIN BARRIER
Custom designed and built to any specifications. Reinforced grommets for hanging. Velcro closure is optional.

LEAD VINYL SHEETING
Available with or without extra strength supported backing. Although our lead sheeting is designed for radiation protection, it's also providing an excellent sound proofing barrier. Available in various thicknesses and widths for custom uses. Contact our office for price quotes.

TILE
Full .5mm equivalency. Lead vinyl tiles with tongue and groove for fast, easy installation and maximum radiation protection.

NOTICE
SELLER'S EXCLUSIVE WARRANTY DISCLAIMER OF ALL IMPLIED WARRANTIES

Seller warrants that seller's products will be free from all defects due to faulty workmanship for a period of one year. Seller's obligation is hereby limited exclusively to repair and replacement of any defective products.

Seller hereby **DISCLAIMS** all other warranties, either expressed or implied including any implied warranty of **MERCHANTIBILITY**. There are no warranties which extend beyond the description on the face hereof.

Courtesy of Shielding, Inc., Madras, Ore.

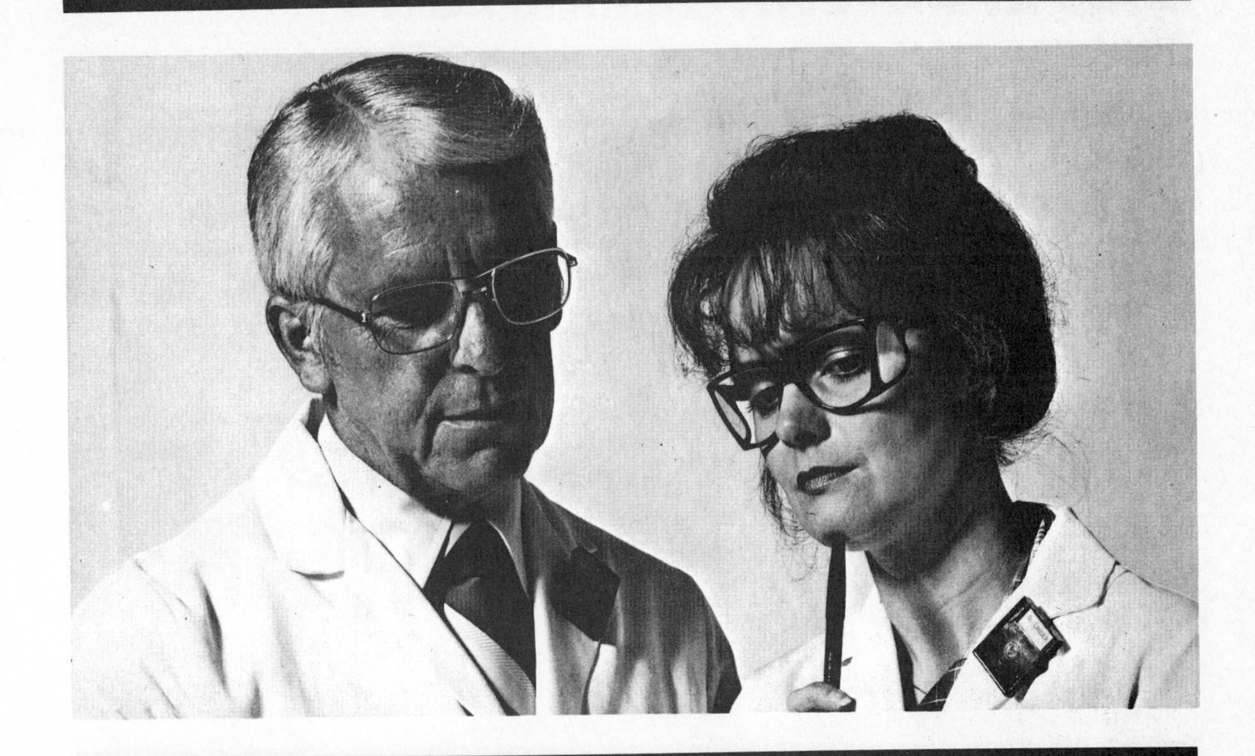

KEEP YOUR EYES SAFE FROM RADIATION...

comfortably and confidently

Shielding, Inc., offers protection for eyes in four formats; standard, wrap-around, contour and clip-ons for regular glasses.

All offer exceptional shielding values combined with lightweight comfort and extremely high optical quality. Lenses provide .75 mm lead equivalency and are anti-reflection coated and tempered to meet FDA impact resistance requirements.

Personnel can now confidently reduce the possibilities of cataracts and still work comfortably without impaired vision. For example, in recent Dose Reduction studies,* Shielding's Wrap Arounds had the highest dose reduction for direct as well as peripheral radiation sources.

*See other side

A. Standard Format Model 50
B. Wrap-Around Model 53
C. Clip Ons Model 51
D. Contour Model 59

*PRESCRIPTIONS
Standard format prescription glasses are Model 55. Wrap around prescription glasses are Model 57. Contour format not available in prescription.

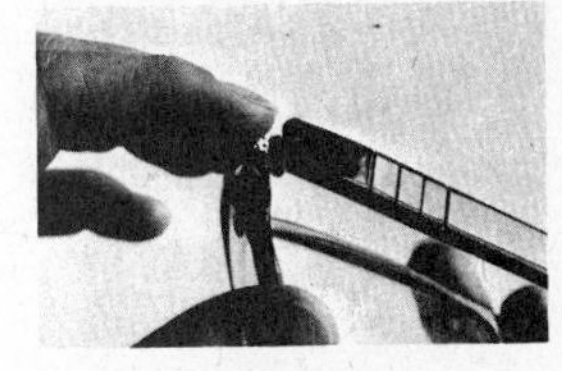

New Design Provides Comfort

Exclusive Op-Tite® adjustable rotors provide tension adjustment for the wearer. The four pins in the temple piece provide additional flex to enhance an exact fit.

® Op-Tite Patent by Lazarus

**For prescription orders, send diopter correction and distance between pupils.*

RADIATION SHIELDING EYEGLASSES

Model	Description	Unit Price
50	Standard Format, non-prescription	$155.00
51	Clip-on	155.00
53	Wrap-Around, non-prescription	197.00
55	Standard Format, prescription	215.00
57	Wrap-Around, prescription	275.00
59	Contour-Format, non-prescription	185.00
B	Add to Prescription Price for Bi-Focals	100.00

Courtesy of Shielding, Inc., Madras, Ore.

COMPARATIVE INFORMATION

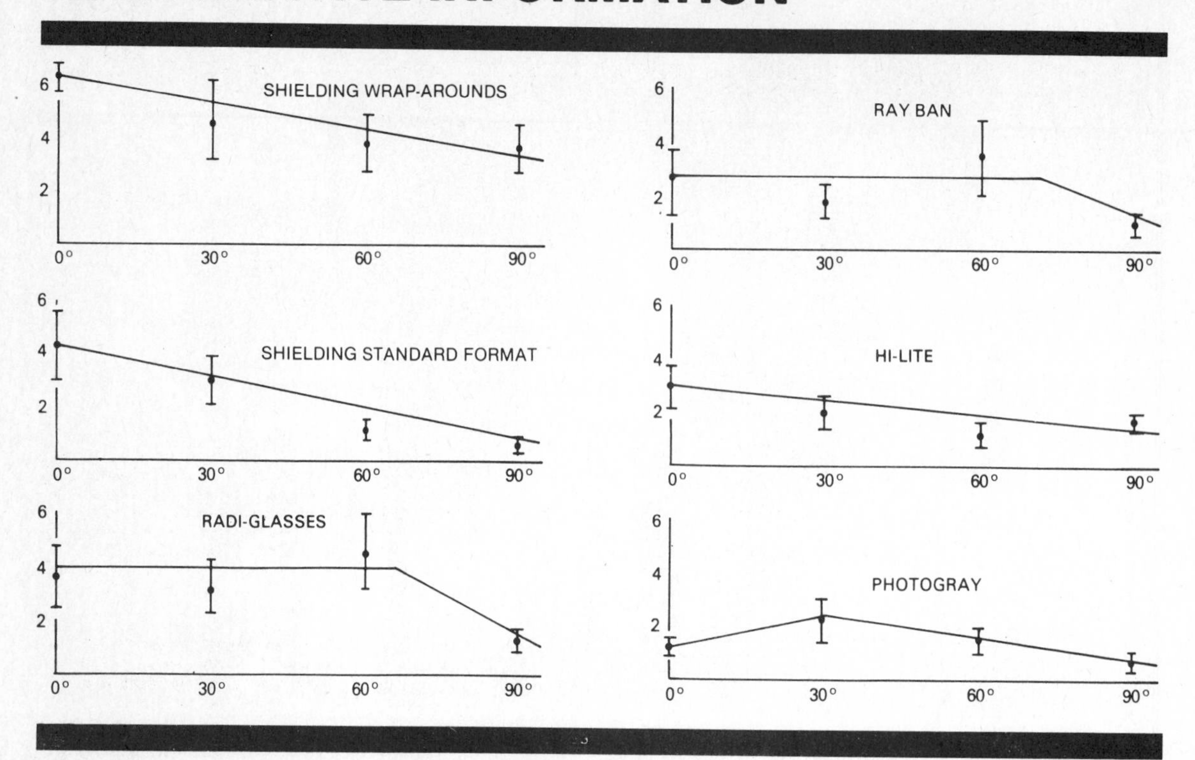

DOSE REDUCTION FACTOR

"...it should be noted that when the (subject's) head is turned, so that it is looking away from the source of the X-rays, the dose radiation factor approaches 1. Thus most glasses offer no protection at 90 degrees from the beam, and may even cause more exposure. The only exception are the Shielding...wrap-arounds."

☐ Source: Test data, University of Washington, School of Medicine and University Hospital Department of Radiology

INCIDENCE OF CATARACTS
Treatment Time—3 weeks to 3 months

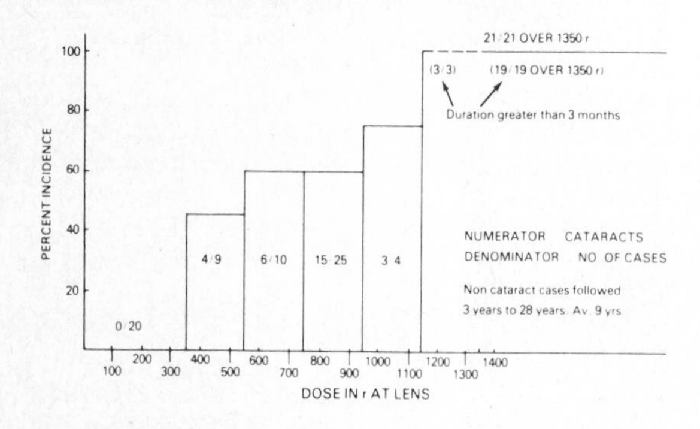

DOSES OF X OR GAMMA RADIATION TO LENS IN 97 CASES OF RADIATION CATARACT AND 70 CASES WITHOUT LENS OPACITIES

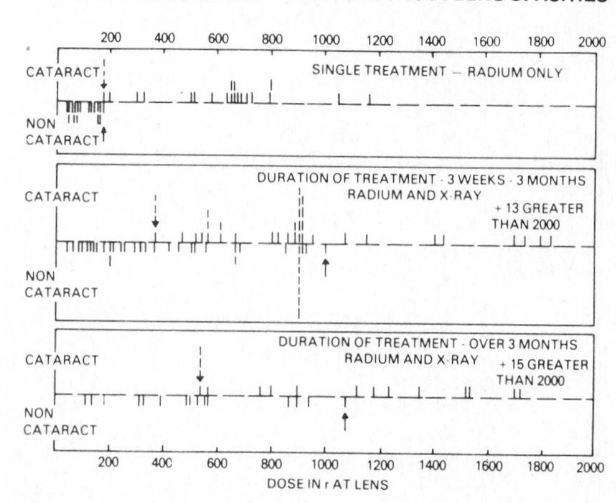

Graph Source:
George R. Merriam, Jr. and Elizabeth F. Focht, Radiology, May, 1967.

Shielding, Inc. P.O. Box 578 • Madras, Oregon 97741

Courtesy of Shielding, Inc., Madras, Ore.

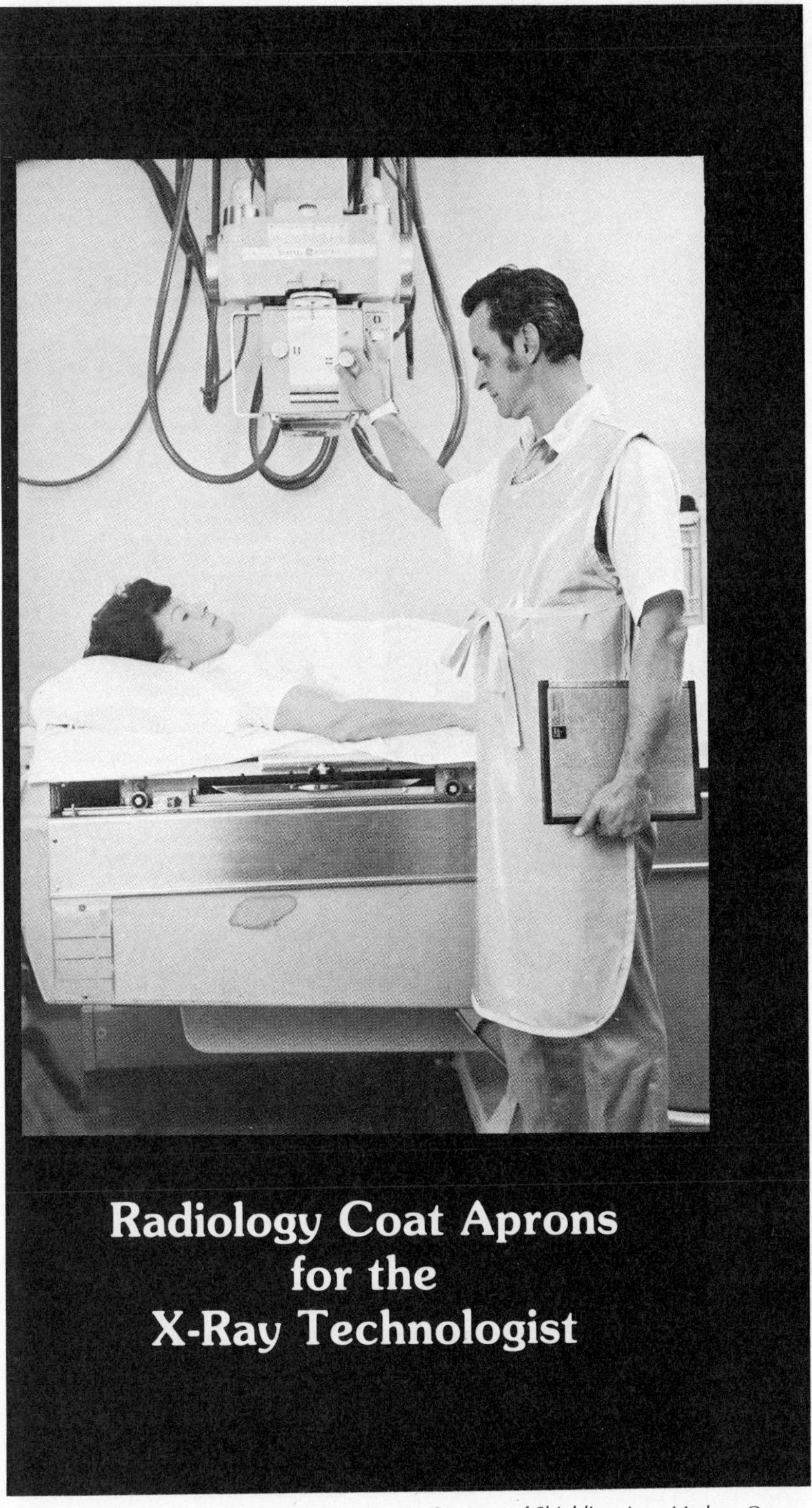

Courtesy of Shielding, Inc., Madras, Ore.

Catalog Sheet No. 4

Full Radiology Protection And Maximum Flexibility In A Light Weight And Full Lead Coat Apron

And lighter in cost, too. Yet with full protection. That's what Shielding's Coat Apron is all about. And that's what adds up to less fatigue for the wearer. Freedom of movement, too, because Shielding's exclusive Multi-Ply construction (a development of Shielding research) is more flexible than competitive materials.

There's no skimping on protection, however. Shielding's Coat Apron is available in a choice of .3mm, .5mm and .75mm equivalency.

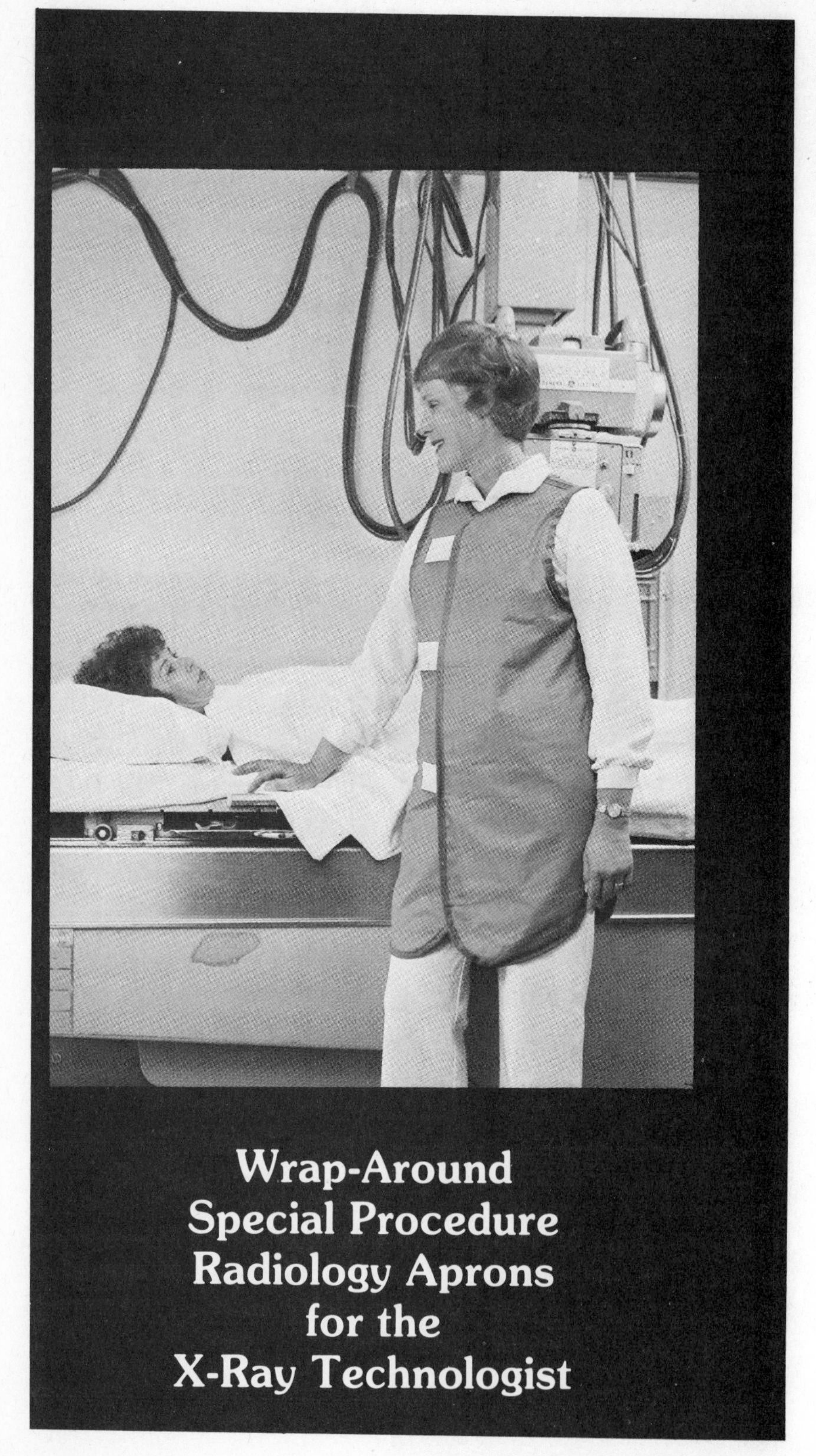

Courtesy of Shielding, Inc., Madras, Ore.

Catalog Sheet No. 2

The Easy-to-Use Wrap-Around Apron When Front and Back Radiology Protection Is Needed

A "Special Procedure Apron" especially designed for front and back protection, Shielding's Wrap-Around offers more protection at lower cost than anything that can be found. Only the nation's leading manufacturer of X-Ray protective garments could turn out so much protection at such low cost.

Like all protective garments by Shielding, the Wrap-Around Apron offers the wearer more freedom of movement; because Shielding's exclusive Multi-Ply construction (a development of Shielding research) is more flexible than competitive materials. Fatigue is reduced due to the extra wide shoulders which distribute the weight over a greater area in addition to providing maximum protection. Closure is easy and positive, because Shielding's Wrap-Around Apron is equipped with easy-to-use Velcro brand hook and loop fasteners at the apron's front. Another Shielding "plus" is its availability with reversible covers — a distinct advantage for multiple users.

Standard radiology protection for Shielding's Wrap-Around Apron is full .5mm equivalency in the front and .3mm in the back, with the option of .5mm front and back. However, they are available with .5mm front protection with no back protection. In this case, the back is cut down for use as an easy-to-don coat style apron without the fuss of straps.

When You Need Special Procedure Protection . . .

. . . consider the extra advantages of Shielding's Wrap-Around Apron in either Multi-Ply or Nu-Liteply construction.

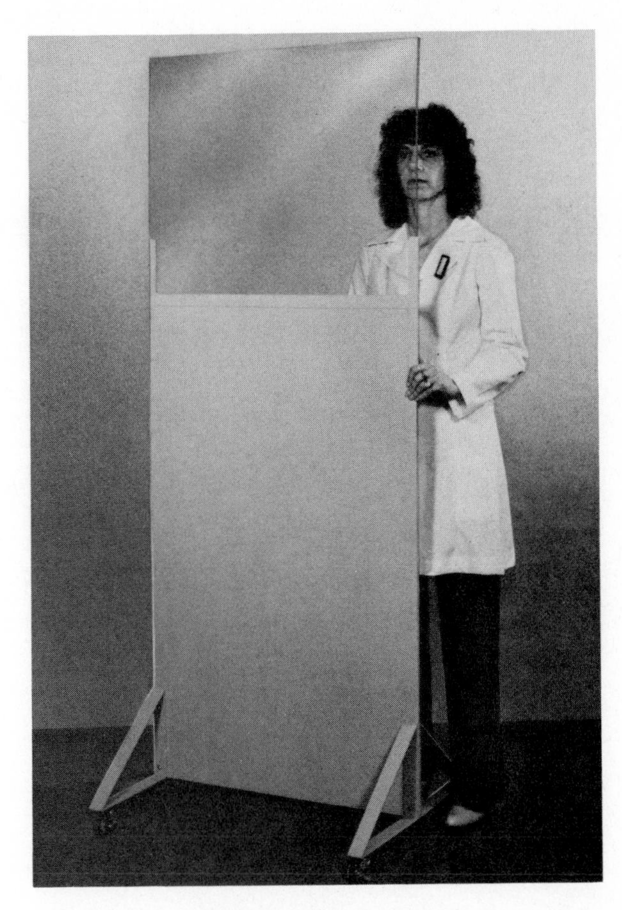

Mobile x-Ray Barrier

Courtesy of Nuclear Associates, Carle Place, N.Y.

RADIATION USE AND SAFETY

ASSESSMENT

1. To become knowledgeable in the care and use of radiographic equipment through in-service education.
2. To ensure patient and staff safety during a radiographic procedure through proper use of shielding and other safety equipment.
3. To educate physicians who are not radiologists in the proper use of radiologic equipment. To emphasize to employees the potential dangers of radiation and the importance of radiation safety.

PLAN

1. To work closely with the radiology department in promoting patient and staff safety during radiologic procedures.
2. To give each employee, through orientation and in-service education, a basic understanding of radiation use, safety, and potential dangers.
3. For nonradiology users of x-ray equipment, an individual or group program in the proper use of the equipment, radiation safety, and applications is essential to obtain the best results from the study.

Implementation	*Rationale and Key Points*
1. Lectures in the use, dangers, and safeguards of radiation	1.1. Lectures will be conducted by the Administrative Manager of x-ray.
2. Special procedure manual includes manufacturer's specifications on the proper use of the equipment	2.1. X-ray equipment is very expensive. Proper use will keep it in the best working order.

Implementation	*Rationale and Key Points*
3. Protection of the staff and patient during the procedure	3.1. Protective equipment: a. Exposure badge b. Lead apron c. Radiation sign on the door d. Face the radiation source and keep as much distance as possible from the direct beam e. Stay within safe fluoro limits (5 min. normal; 10 min. upper limits; over 20 min., ask for assistance).
4. Protection of the patient during examination	4.1. Follow these procedures: a. Check identification b. Check allergic history c. If patient is a female, ask about pregnancy d. Test I.V. for patency. e. Patient under the age of 40: cover gonads if test allows f. Have emergency cart and/or drugs in room g. Stay within time limits for radiation.
5. Care of auxiliary equipment	5.1. Follow these procedures: a. Wear exposure badge in every case, but do not take it home b. At the end of the case, hang up the lead aprons c. When not making exposures, radiation machine should be in the "off" position d. If direct readout dosimeters are used, record the readings and return dosimeter to storage area.

APPROVED: ____________________

DATE: ____________________

RADIATION BADGES

SUBJECT: Radiation Badges

RULES

1. Radiation badges will be worn by all employees assisting on a regular basis with x-ray procedures that utilize ionizing radiation.
   a. All employees in need of a radiation badge will obtain an application form from the x-ray department.
   b. Radiation badge films will be changed monthly.
   c. Direct readout dosimeters will be read prior to each use and after each use. The amount of radiation received will be recorded in a log book.
   d. The chief technologist will notify employees if excessive amounts of radiation exposure occur.
2. The radiation badges will be left in the department when not in use.
   a. There will be a rack for radiation badges away from any source of radiation.
   b. Leaving badges in the department will provide control over extemporaneous exposure (sunshine), prevent loss of the badge, and facilitate monthly exchange of badges.

Responsibility	*Action*
1. Individual responsibility of team members working with ionizing radiation.	1.1. Request film badges from x-ray department. 1.2. Change the film badges monthly. 1.3. Remove the badge before leaving the hospital.
2. Chief Technologist or Administrative Assistant of x-ray Department.	2.1. Keeps accurate records of radition exposure. 2.2. Notifies employees of excessive exposure.

APPROVED: ______________________

DATE: ______________________

PERSONAL DOSIMETER REQUEST FORM

The following information is required for the assignment of a personal dosimeter to ______________________________ for radiation monitoring.

Full Name ______________________________ Date ______________

Birth Date ______________________________ S.S. No. ______________

Department ______________________ Hospital Phone No. ______________

Supervisor ______________________ Staff ________ Student ________

Type of radiation to which you will be exposed:

X-ray ______________________ Diagnostic ______________________

Therapeutic ______________________ Other (specify) ______________________

____

Beta-gamma ______________________ Radioisotopes to be used ______________

Have you been previously exposed to radiation at another place of employment?

Yes ______________ No ______________

If Yes, please give place or places and circumstances.

____

____

Date monitoring is to begin ______________________________

Date dosimeter was issued ______________________________

FOR RADIATION SAFETY OFFICER OR CHIEF TECHNOLOGIST

X-ray ______________ Weekly ______________ Quarterly ______________

Beta ______________ Biweekly ______________ Dosimeter no. ______________

Gamma ______________ Other ______________

Approved: ______________________________
Radiation Therapy Officer/Chief Technologist

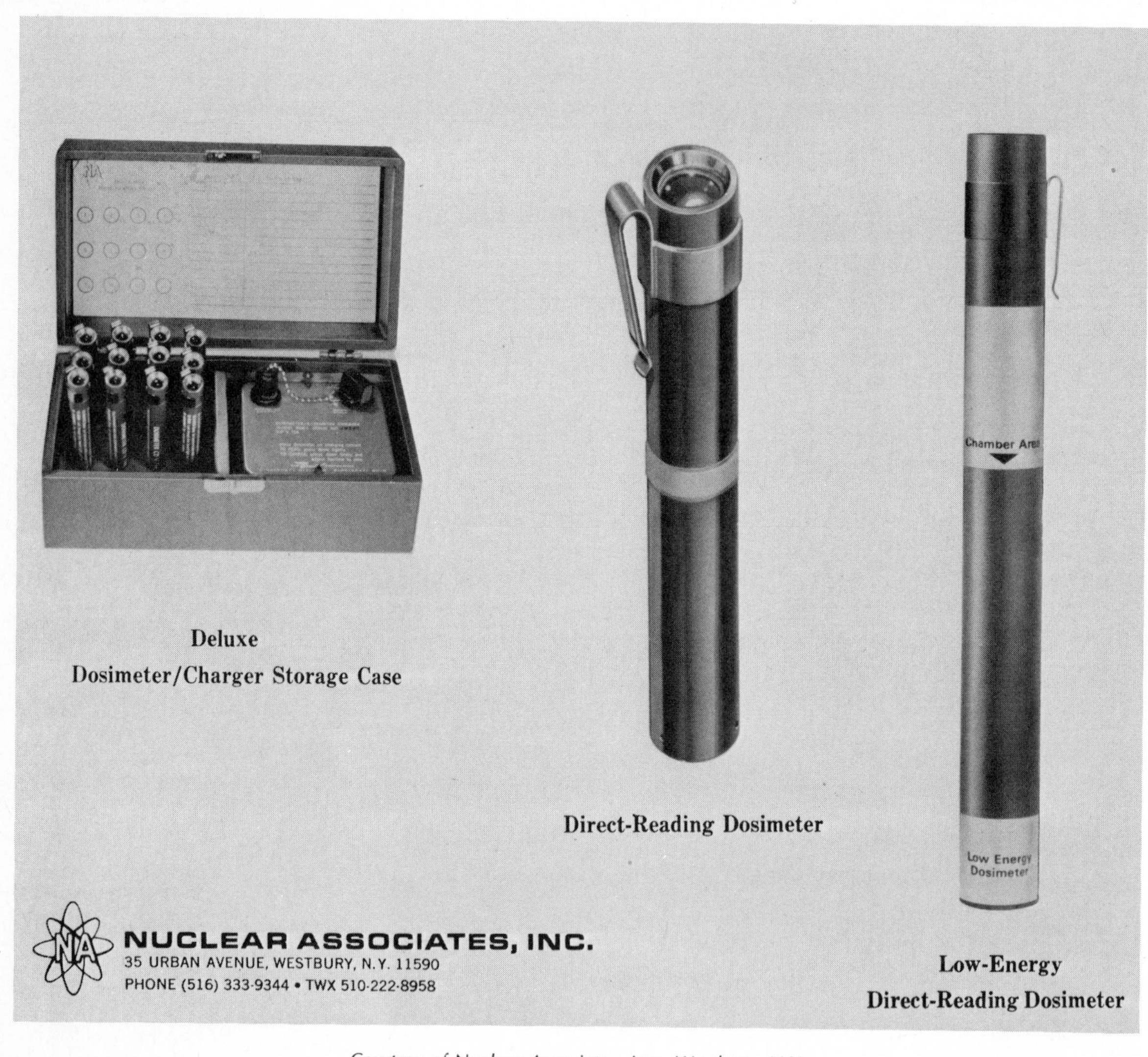

Courtesy of Nuclear Associates, Inc., Westbury, N.Y.

DIRECT READOUT DOSIMETER LOG

Date	Name	Starting Milli R's	End Milli R's
1.			
2.			
3.			
4.			
5.			
6.			
7.			
8.			
9.			
10.			
11.			
12.			
13.			
14.			
15.			
16.			
17.			
18.			
19.			
20.			
21.			
22.			
23.			
24.			
25.			

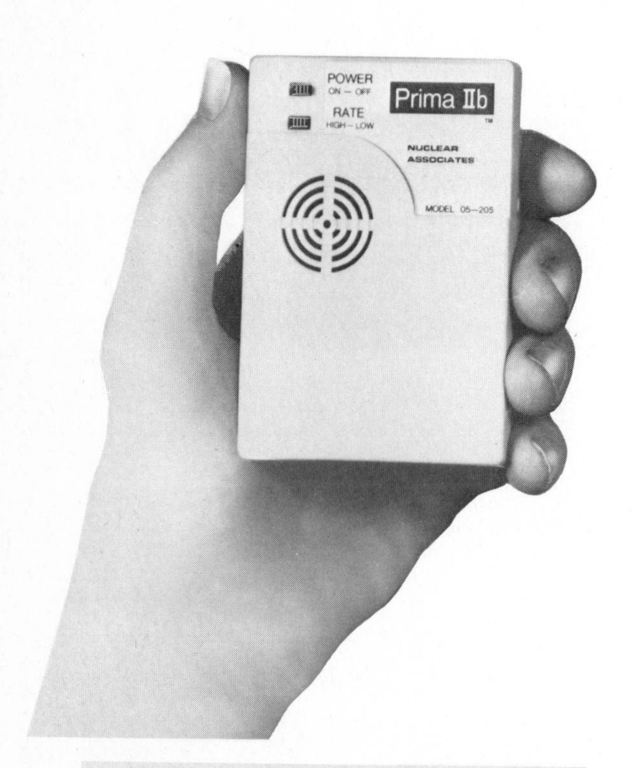

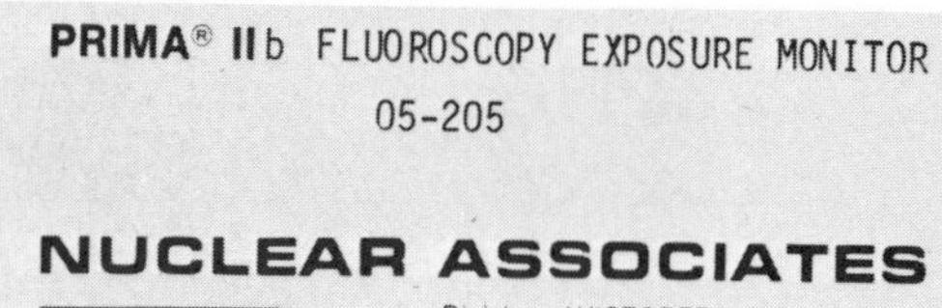

Courtesy of Nuclear Associates, Carle Place, N.Y.

TESTING OF PROTECTIVE EQUIPMENT

SUBJECT: Lead Aprons

RULES

1. Lead aprons in the ______________________ x-ray department should be checked every six months for rips in the material and cracks in the lead. The results will be noted in an equipment log.
2. If the aprons examined show cracks in the lead, they will be taken out of service and replaced.
3. The criteria for the examination of the aprons shall include:
   a. *General Appearance*
   (1) Ties all present and intact
   (2) No tears or rips in vinyl coverings
   b. *Radiographic Examination*
   (1) Use 80 kV and 1 mAs
   (2) Check for any breaks in the lead
   (3) Record in the equipment log
   (4) Report aprons that are faulty and take out of circulation.

Responsibility	*Action*
1. Any technologist in the X-ray department	1.1. Aprons checked every six months and logged. 1.2. Cracked aprons taken out of circulation and reported to chief technologist.

Approved: ______________________
Director of Radiology

Date: ______________________

EQUIPMENT LOG: LEAD APRONS AND GLOVES

Apron No.	Date	General Appearance	Comments

TECHNOLOGIST–SERVICE ENGINEER RELATIONSHIP

Overview

It is necessary to nurture a relationship between the technologists and the x-ray service engineers if the radiographic quality control program is to be successful and to run smoothly. This relationship, like that of the radiologist and technologist, is an important relationship to the radiographic department.

TECHNOLOGIST DUTIES

1. *Documentation:* It is very important that the technologist document for the service engineers the following parameters:
   a. What was the study being done when the problem occurred?
   b. What technique was used in this procedure?
   c. If an exposure was made, **save the film.** Often when a radiograph is made and it is of poor physical image quality, it is thrown away. The film is often the key to what the problem was and also a starting place for troubleshooting the x-ray system.
   d. Was the x-ray system used on previous cases and if so what did the films look like?
   e. If the technologist attempted to correct the problem, what parameters were changed and what was the result of these changes?
   f. Did any other variables enter into the picture, such as processing or using a different type of film?
   g. Does the problem seem to be intermittent?
   h. Has the problem occurred in the near past? What was done to correct the problem?
2. *Reporting the problem:* Unfortunately, many minor mechanical and/or electronic problems are not reported so that appropriate action can be taken. It is much easier to correct small problems with no downtime for the x-ray system than large problems, which may take the room out of operation for most of a workday.

SERVICE ENGINEER DUTIES

1. Talk to the technologist who experienced the problem. Often information obtained from other sources is inaccurate.
2. *Effective communication:* One of the greatest problems between technologists and service engineers is in defining what the problem is. Technologists are not used to communicating in electronic terms and this leads to misunderstanding about the problem. This roadblock, however, can be removed by questioning the technologist or asking him or her to repeat what was done during the case to demonstrate the problem.
3. If a radiograph was made, look at it.
4. Troubleshoot the problem.

ESTABLISHMENT OF TECHNIQUE CHARTS FOR RADIOGRAPHIC X-RAY SYSTEMS

Overview

It is essential for quality control and a quality assurance program that technique charts be established for each radiographic x-ray system and procedures done on those x-ray systems. It is through these charts that reproducibility of good radiographs is established throughout the department.

Film companies often will give technique charts to the hospital when they use their supplies. Dupont has an excellent representation of such technique charts. (See Appendix)

TECHNIQUE DISCIPLINES USED IN ESTABLISHMENT OF TECHNIQUE CHARTS IN THE UNITED STATES

TECHNIQUE DISCIPLINES

1. *High kilovoltage* (100–150 kVp): This discipline is founded upon fixed voltage techniques in the high kilovolt peak (kVp) ranges. The radiographic studies most often done by this technique are chest, barium studies, prenatal studies, and lumbar spines.
2. *Variable kilovoltage:* This is a manual discipline where milliampere-seconds (mAs) remain constant and the kVp changes per density and size of the anatomical part of the body being radiographed. The average increase is 2 kVp/cm of body thickness (density).
3. *Fixed kilovoltage:* In this discipline the kVp remains constant and the mAs varies to compensate for the size of the patient. The kVp is usually adequate to penetrate the anatomical body part. The result is a radiograph of a suitable scale of contrast.
4. *Automated:* This technique produces films of optimum quality without the requirement of body measurements. This technique lends itself well to the falling load and/or constant potential generator. In this discipline, an ionization chamber or a photocell is used in the termination of the exposure once sufficient radiation has reached its receptor.
5. *Body habitus:* This is the most difficult discipline in that it requires the body typing of patients prior to measuring the anatomical part and consulting one of four technique charts. The kVp and mAs vary to balance out radiographic contrast.

RADIOGRAPHIC TECHNIC CALCULATOR*

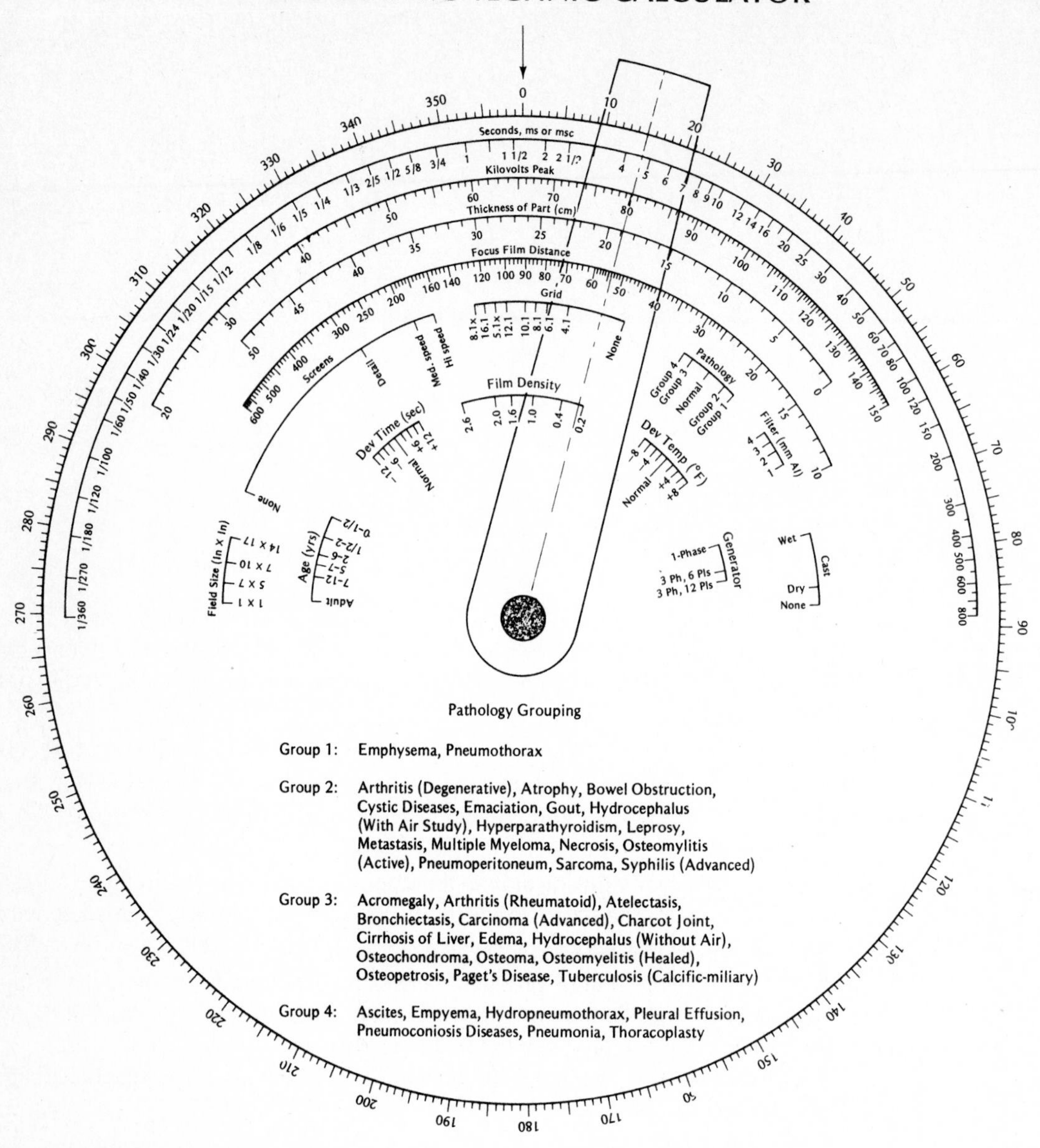

Examples of Use

Problem 1: At 30 inches FFD, mas required is 20. What mas would you use at 60 inches?

Solution: Set cursor at arrow on base plate. Hold cursor in place with the left hand. Rotate wheel so that the setting "30" on FFD scale lines up with cursor. Hold wheel in place with right hand. Move cursor to setting "60" on FFD scale. Now rotate wheel and make setting "20" on the mas scale line up with cursor. Bring cursor back to arrow and read on mas scale. The answer should be 80 mas.

**The radiographic technic calculator was designed by Dr. Gopala Rao, Assistant Professor of Radiology (Physics), The Johns Hopkins Hospital, Baltimore, Md., Copyright 1970, and is reproduced with permission.*

Problem 2:

Old Factors	New Factors
80 KVP	70 KVP
30 cm part	20 cm part
No grid	8:1 grid
20 mas	What is the new mas?

Solution:

Set cursor against arrow on base plate.
Move wheel to 80 KVP
Move cursor to 70 KVP
Move wheel to 30 cm part
Move cursor to 20 cm part
Move wheel to "no grid"
Move cursor to "8:1 grid"
Move wheel to 20 mas
Move cursor to arrow and read on mas scale
The answer should be 33 mas.

Problem 3:

Old Factors	New Factors
40 mas	20 mas
No grid	8:1 grid
85 KVP	What is the new KVP?

Solution: The answer is 126 KVP.

Problem 4:

Old Factors	New Factors
50 KVP	60 KVP
No cast	Wet cast
No screens	Detail screens
One second	What is the new time?

Solution: The answer is 1/10 second.

Remember that you move the wheel to set old factors and the cursor to set new factors.

TYPICAL RADIOGRAPHIC TECHNIQUES

		Approx. MAS for Thickness Specified		
Body Part	*KVP*	*Thickness (cm)*	*MAS*	*Other Factors Assumed*
Extremities	55	10	125	No screens, no grid and 40" FFD
Extremities	55	10	5	Med. screens, no grid and 40" FFD
Trunk of body (AP)	75*	20	40	
Trunk of body (Lat)	85	30	200	
Thoracic spine (AP)	75	20	75	
Thoracic spine (Obl and Lat)	85	30	75	
Cervical spine (AP)	65	15	50	
Cervical spine (Obl and Lat)	70	10	15	Med. screens, 8:1 grid and 40" FFD
Skull (AP, PA, Townes)	75	20	40	
Skull (Lat)	60	15	40	
Skull (Basal)	90	15	40	
Sinuses (PA 0°, Basal)	75	20	75	
Sinuses (PA 15°)	70	20	75	
Sinuses (Lat)	65	15	20	
Chest (PA)	75	25	10	
Chest (Obl)	85	30	30	Med. screens, no grid and 72" FFD
Chest (Lat)	85	30	40	

*Add 10 KVP if contrast medium is present.

RADIOGRAPHIC TECHNIC CONVERSION GUIDE

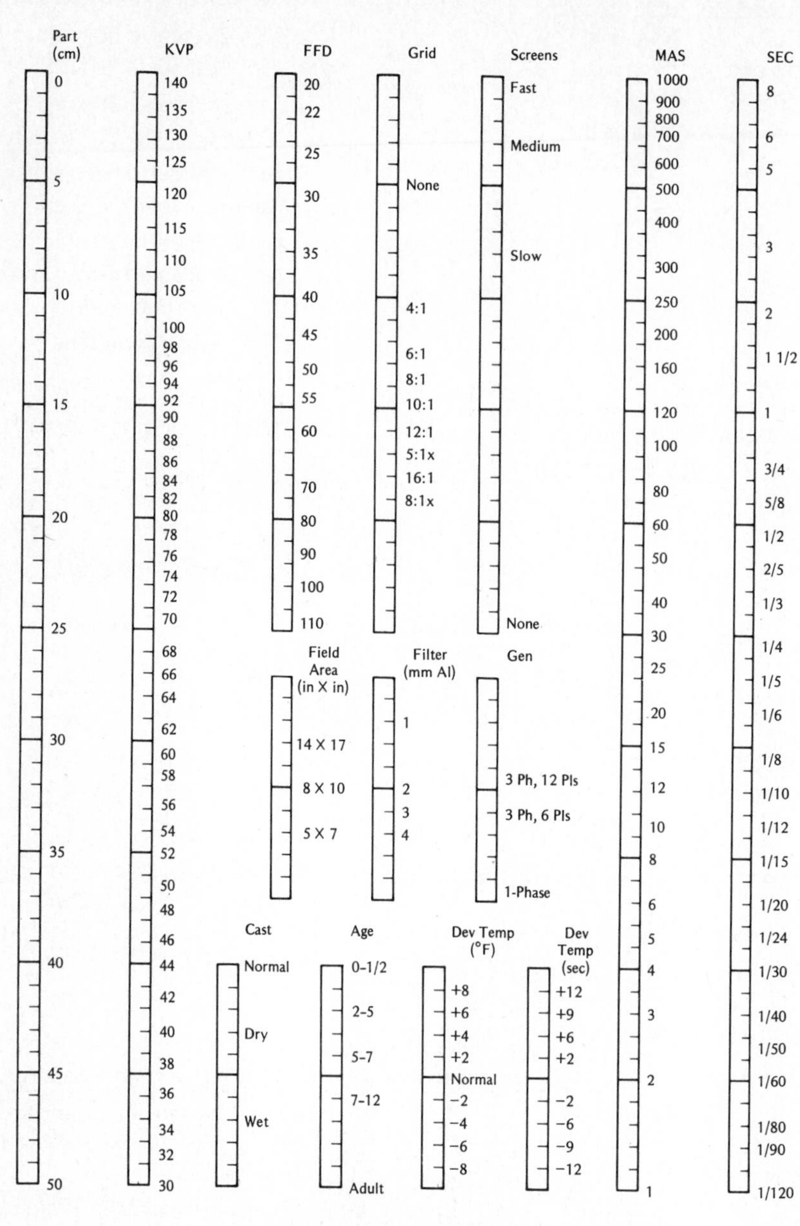

Examples of Use

Problem 1: If a 25 cm patient requires 1/10 second exposure, what will a 35 cm patient require?

Solution: Locate the numbers 25 and 35 on the "Part" scale. Count the number of scale divisions between the two. In this case, you will find that you have to go down 10 scale divisions to move from 25 and 35.

Problem 2: The Lateral Cervical Spine is usually taken at 72 inches FFD with a 1/4 second exposure. The patient cannot sit up: you have to make the exposure with the patient lying down. You do not wish to deviate very much from the usual 72 inches; so you have selected an FFD of 48 inches. What should be your new time?

Solution: Locate the numbers 72 and 48 on the FFD scale. Count the number of scale divisions between the two. In this case, you will find that you have top go up about 6 scale divisions to move from 72 to 48.

Now locate 1/4 second on "Second" scale. Go down 6 divisions. Read the answer (1/6 sec.)

ELIMINATE THE NEED TO MAKE OR REVISE WALL CHARTS.

ACHIEVE HIGHEST DEPARTMENTAL QUALITY STANDARDS AND PROCEDURES

SUPERTECH PATENTED TECHNIQUE CHART ADJUSTS TO EVERY MACHINE.

QUICK REFERENCE BOOK OF EVERYTHING PERTINENT TO THE PROCEDURE.

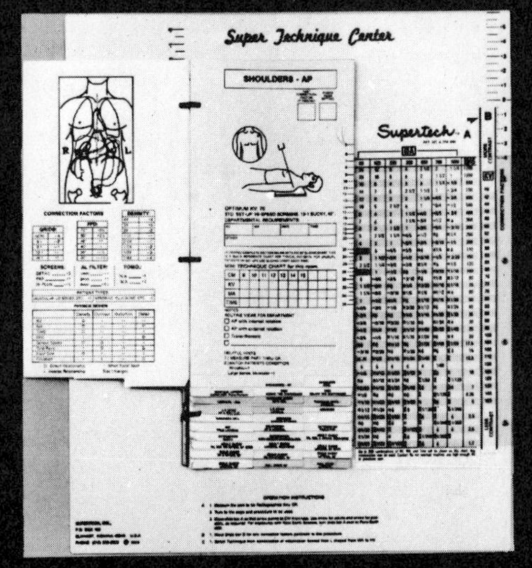

MOUNTS ON WALL BESIDE EACH CENTRAL PANEL

Contact

Supertech, Inc.
P.O. Box 186
Elkhart, IN 46515
219/293-2832

YOUR LOCAL DEALER

Courtesy of Supertech, Inc., Elkhart, Ind.

SPECIFIC AREAS TO BE MONITORED BY QUALITY CONTROL IN THE DEPARTMENT OF RADIOLOGY

Overview

The specific parameters that are to be monitored in a diagnostic radiology facility should be determined by the quality control committee of that facility. The decision should be based upon an analysis conducted in that facility of the expected benefits and cost of such a program.

Other factors to be taken into consideration are the physical size of the facility, the number of cases, the type of procedures, and the resources of the radiographic facility. Before initiating a program, the radiographic quality control committee may gain insight into their needs by visiting a similar facility with a functioning program.

NINE BASIC AREAS OF RADIOGRAPHIC MONITORING

1. Basic acceptable performance characteristics of the x-ray system
2. Radiographic film processing
3. Physical image quality
4. Film-screen characteristics
5. Cassettes and grids
6. Film
7. Darkroom and safe lights
8. View boxes, hot lamps, projectors
9. Technical error

PARAMETERS USED IN THE ESTABLISHMENT OF OPERATING STANDARDS FOR QUALITY CONTROL

1. General radiographic x-ray system:
    - a. Linearity and reproducibility of mA stations.
    - b. Accuracy and reproducibility of kVp and timer stations.
    - c. Reproducibility of radiation output.
    - d. Consistency of focal spot size.
    - e. Accuracy of distance indicators from the radiation source to the film.
    - f. Light/x-ray film congruence (film-screen congruence).
    - g. Half-value layer.
    - h. Chart of the representative entrance skin radiation exposure.
2. Fluoroscopic x-ray system:
    - a. mA, kVp, and exposure time must be accurate and reproducible.
    - b. Radiation output must be reproducible.
    - c. Consistency of focal spot size, pin hole camera.
    - d. Radiation exposure rates at the table top.
    - e. Precision of mechanism for centering alignment.
    - f. Collimation must meet state and federal guidelines.
    - g. Half-value layer.
    - h. Representative entrance skin exposures.
3. Generator: automatic exposure devices (phototiming):
    - a. Reproducibility
    - b. Field sensitivity matching
    - c. kVp compensation
    - d. Backup timer verification
    - e. Minimum response time
4. Image intensifier system:
    - a. Resolution: sharpness, expressed in line pairs per millimeter
    - b. Low contrast performance
    - c. Glare and distortion
    - d. Focusing
    - e. Physical alignment of camera and collimating lenses

5. Physical image quality:
   a. *Large Area Contrast:* this is represented by the characteristic H and D curve (film-screen system). An H and D curve is a graphical expression of the manner in which optical density in the process varies with the radiation exposure.
   b. *Intrinsic screen contrast:* this is the inherent phosphor contrast of the screen. It is accumulative with the subject contrast-radiation and the image receptor, film, and contrast.
   c. *Detail contrast:* sharpness of the boundary limits the overall contrast when the test object is reduced. The subject radiation, however, remains the same.
   d. *Noise contrast: quantum mottle:* the noise contrast is a function of the large area contrast, detail contrast, and the number of quanta used in the film's production.
   e. *Non-image-forming radiation:*
      (1) Scatter from patient
      (2) Off-focus radiation
      (3) Intensified fluoroscopy non-image-forming radiation

These five areas will be checked by a physicist to determine the values indirectly or by the service engineer to measure the parameters directly.

6. Radiographic and cineradiographic film processor:
   a. An index of speed: radiation dose
   b. An index of contrast: contrast, noise, resolution
   c. Base plus fog: clearest portion of film
   d. Temperature: developer, water, dryer
   e. Chemical activity: agitation
   f. Replenishment rate
   g. Transport speed
   h. Amount of film processed over a specified amount of time
   i. Film artifact identification
7. Properties of radiographic and intensifying screens that influence the effectiveness of a film-screen combination:
   a. *Speed:* reciprocal of exposure in roentgens required to produce a net optical density of 1 in the processed film
   b. *Contrast factor:* average gradient, gamma
   c. *Latitude:* this is the range of exposures over which an x-ray may be obtained
8. Film-screen combination facts:
   a. *Speed:* radiation dose
   b. *Film-screen congruence:* combination
   c. *Image quality:* diagnostic outcome
   d. *Required image quality:* procedure asked for
   e. *Miscellaneous factors other than film-screen combination affecting image quality*
   f. *Physical image quality:* a combination of contrast, resolution, and noise
   g. *Variables: diagnostic outcome*
      (1) Radiologist preference
      (2) Individuality of patient
      (3) Medical decisions
      (4) Miscellaneous

9. Film:
    a. *More than one type of film used:* need two quality control programs
    b. *Master control film:* emulsion batch number
    c. *Cross over test:* changing emulsion lot numbers
    d. *Storage and handling:* cool, dark, away from radiation or light source
    e. *Amount and types of film on hand*
10. Cassettes:
    a. General age and condition of the holder (cassette)
    b. General age and condition of the screens
    c. Film-screen contact
    d. Light leaks
    e. Artifact identification
11. Grids:
    a. Alignment and focal distance (system geometry)
    b. Identification of artifacts
12. Darkroom:
    a. Light leaks
    b. Safe lights
    c. Indicator lights
13. View boxes, hot lights, projectors:
    a. Consistency of light output with time
    b. Consistency of light between one view box and another
    c. General surface conditions
    d. Cleaning

The parameters discussed from number 6 to 13 may be monitored by the quality control coodinator, service engineers, physicist, or any combination of the three.

APPROVED: ____________________

DATE: ____________________

PARAMETERS USED IN THE DETERMINATION OF QUALITY CONTROL FREQUENCY CHECKS

1. X-ray equipment:
   a. *Age of the equipment:* as a general rule, the older the equipment, the more often it must be checked.
   b. *Volume of work:* the greater the patient volume, the more often the room needs preventive maintenance.
   c. *Complexity of the equipment:* the more complex the equipment, the greater the frequency checks will be.
   d. *Technologist expertise:* the greater the knowledge of the technologist in the operation of the equipment and the care of the equipment, the less problems will be encountered.
   e. *Critical areas:* the more critical the area, the more frequent the checks. These areas include x-ray special procedures, cardiac catheterization, and computerized axial tomography, to name a few.
2. Generators, radiographic:
   a. Complex: fluoroscopic radiographic x-ray system:
      (1) Three-phase generator:
         (a) Falling load
         (b) Constant potential
         (c) 6- or 12-pulse, nonfalling load, 3-point technique
      (2) Phototiming: automatic timing device

      These generators need more preventive maintenance due to their complexity and their use.
   b. Simple: straight radiographic x-ray system:
      (1) Single phase, 2 pulse
      (2) No phototiming, automatic timing device
3. Radiographic processors:
   a. *Age of the processor:* the older the processor, the more preventive maintenance it will need.
   b. *Volume of work:* the greater the volume of work, the more often the processor will need to be cleaned and have preventive maintenance.
   c. *Replenishment rate:* the replenishment rate will depend upon the volume of work and the type of film used.

d. *Temperature of the developer, water, and dryer:* these parameters will depend upon the manufacturer of the equipment and the type of film.

4. Darkroom integrity: The darkroom and safe lights should be checked at a minimum of every three months or more often as the need arises.
5. View boxes, hot lights, and projectors: These items need to be cleaned and preventive maintenance done every three months. Cleaning the surface of the items needs to be done daily.

RADIOGRAPHIC SUPPLIES AND MANUFACTURERS

Intensifying Screens	*Automatic Film Processing*	*Automatic Film Processors*	*Photographic Chemicals*
Du Pont	DuPont	Dupont	DuPont
Detail	Cronex 4	QC-1	Kodak
Fast Detail	Cronex 6		Picker
Par	Cronex 6 Plus	Kodak	Ilford
Hi Speed	Cronex 7	M6AW	Gafmed
Hi Plus	Cronex 8	M7A	AGFA-Gevaert
Lightening Plus	Lo Dose	RP X-Omat	Others
Quanta II			
Lo Dose I	Kodak	Pako	
Lo Dose II	XG	17X - 90	
Quanta III	XRP	17X - 2	
Quanta V	XL	17X - CW	
	XR		
Kodak	XS	Litton (4)	
X-Omatic Fine	Ortho-G		
X-Omatic Regular	Min-R	General Electric (3)	
Lanex Regular	Blue Brand		
Min-R		AGFA-Gevaert (1)	
	3M		
3M	Alpha 8		
Alpha 8	Alpha 4		
Alpha 4	Alpha M		
Alpha M			
	AGFA-Gevaert		
AGFA-Gevaert	Others		
Picker			
Seamens			
Ilford			
Gafmed			
Intensi			
Trimax			
Others			

CINE RADIOGRAPHIC SUPPLIES AND MANUFACTURERS

Cine Films	*Automatic Film Cine Processors*	*Cine Photographic Chemicals*	*Cine Projectors*
Kodak	Jamerson	Kodak	Tagarno
CFA	Pako	Du Pont	Vanguard
CFH	Combilabor	Ilford	Others
CFR	Fisher	Many smaller companies	
CFS	Mark		
CFX	Houston-Fearless		
Ilford			
Cinegram F			
DuPont			
Cronex			
AGFA-Gevaert			
Cinefluorography			
Scopix RPI			

BASIC EQUIPMENT AND MATERIALS NEEDED FOR THE RADIOGRAPHIC QUALITY ASSURANCE PROGRAM

I. PROCESSOR QUALITY ASSURANCE

A. Objective

To monitor the radiographic processors and to maintain them within the established control standards set up by the quality control committee of the facility.

B. Equipment

1. *Sensitometer:* This is an instrument containing a light source and a timing mechanism designed to give a reproducible and accurate graded exposure scale to the radiograph.
2. *Pre-exposed sensitometric control monitoring strip:* These strips are supplied by the manufacturer to be used as reference strips for sensitometric (sensi) strips.
3. *Sensitometric control monitoring strip:* This is a piece of radiographic film that has been exposed to a sensitometer or a step wedge and exposed to radiation resulting in a number of densities (shades of gray) post radiographic processing.
4. *Sensitometer step tablets:* These are component parts of a sensitometer that vary the density levels.
5. *Densitometer:* An instrument that accurately measures the densities of an exposed, processed, and dried radiograph.
   a. *Electronic densitometer:* This densitometer is higher priced than visual densitometers and is much more accurate. It should have the capacity of reading densities from 0.0 to 4.0, reproducibility should be ±0.02 and aperture should be of 1 to 3 mm in diameter.
   b. *Visual densitometer:* The visual densitometer (often referred to as the "grease spot" densitometer) is an instrument in which the human eye is used to compare the densities. This is a limiting feature in that the eye can perceive only a small or limited range of the gray scale.
6. *Thermometers:* There are three important aspects to consider when working with thermometers. The first aspect is the accuracy or precision of the instrument. One of the most frequently used thermometers is the dial type with a 6 to 8 inch probe, and a scale of 25° to 125°F. The second aspect of concern is that

a thermometer which contains mercury never should be used in radiographic processing. Mercury is a photographic contaminate even at extremely low levels (parts per million). The third and final aspect is that the thermometer should always be wiped clean to avoid contamination of the chemistries when transferring it from one tank to another.

7. *Photographic (radiographic) control emulsions:* A radiographic control emulsion is a batch of radiographic film with the same manufacturer lot number. A sufficient amount of film should be obtained at one time for a 3 to 6 month period of time.
8. *Control charts:* This is a graphical means of documenting (recording) radiographic processor data.
9. *Silver reclamation and test materials:* Silver reclamation is a substantial means of cost containment within the radiographic department. Most diagnostic facilities can reclaim enough silver to pay for the total cost of the fixer used within the x-ray department. All manufacturer types of silver reclamation devices need to be monitored by quality control. They will require periodic examination and preventive maintenance to ensure that the maximum amount of silver is being reclaimed from the spent fixer. There is never a 100% silver recovery from any type of silver reclamation unit.
10. *Residual fixer test materials:* This procedure is very important in the field of radiology for several reasons. Without testing this parameter, the radiographs may lose the permanency of the radiographic image or it may fade and stain with age. The procedure for doing this test is relatively simple. Test solutions may be obtained directly from the film manufacturer. This should include a series of colored patches indicating the amount of test solution and the stain that should be imparted to the radiograph.
11. *Darkroom interlock system:* Lok-A-Bin system keeps the film bins locked at any time when the access doors are open or overhead lights are on inside the darkroom. At any time when a film is being processed or the film bins are open, it prevents the white lights from coming on. This is an important piece of equipment for cost containment. Losing a full bin of radiographs is costly.
12. *Hydrometer for specific gravity:* Specific gravity can be used to determine the saturation (overconcentration) or dilution (underconcentration) of mixed chemistry.
13. *Miscellaneous equipment:* There are many small items that the quality control coordinator will need to fulfill his or her duties, such as scissors, pens, pencils, rulers, and the like.

II. X-RAY SYSTEM: QUALITY ASSURANCE

A. Objective

To identify the need for corrective action or preventive maintenance in order that a radiographic facility may continue to produce high-quality radiographs.

B. Measurements and Equipment

1. *kVp:* Improperly calibrated x-ray generator leads to kVp fluctuations resulting in radiographs of poor overall density and changes in radiographic contrast.
   a. *Test devices:* The test items are designed to give a repeatable approximation

Deluxe Digital kVp Meter
Courtesy of Nuclear Associates, Inc., Carle Place, N.Y.

of tube potential. Quality control testing will indicate need for corrective maintenance.

(1) Kilovoltage calibrator
(2) Radiographic test cassette: numerous types

b. *Kits and multiple parameter devices:*

(1) Dyanalyzer II
(2) Hivex II
(3) Basic Quality Assurance Program Test Kit
(4) Others

2. *Timer and mAs:* There has been a sophistication in generator timing devices leading to millisecond (ms) exposure times in the last few years. Small timing problems in the shorter time stations result in poor overall density and poor physical quality radiographs. The following test items are used to test the precision and repeatable or reproducible indications for length of exposure. Through quality control testing, indications for corrective action to adjust the timer stations are identified.

a. Test Equipment: timers

(1) Digital X-ray Generator Timer: Measures the timer precision of half-wave to full-wave, or three-phase x-ray generators.

(2) Multiphase spin top: Used in the calibration of constant potential equipment as well as single-phase generators.

(3) Spin top: Used for the calibration of single-phase generators.

b. Test equipment: timers, mAs

(1) Timing and mAs test kit: May be used to test mAs uniformity and precision of the timer in three-phase and constant potential generators.

Digital X-Ray Generator Timer
Courtesy of Nuclear Associates, Inc., Carle Place, N.Y.

c. Other test kits and multiple parameter devices:
   (1) Wisconsin Timing and mAs Kit
   (2) X-ray Pulse Counter
   (3) Dyanalyzer II
   (4) Basic Quality Assurance Text Kit
   (5) Others

3. *Phantoms:* These are man-made devices used in optimizing x-ray systems, training purposes, quality assurance performance evaluations, operating techniques, and research and evaluation of new radiographic products. The phantom simulates a patient.
   a. Types of phantoms:
      (1) Catphen tm: Computerized Axial Tomography
      (2) Fluoroscopic
      (3) Mammographic
      (4) Phantom Patient
      (5) 3M Anatomical Phantoms
      (6) 3M Tomographic Phantoms
      (7) Others

4. *Exposure and exposure rate measurement:* This is one of the most important aspects of radiation safety for the patient, physician, and staff.
   a. Why exposure measurements are necessary:
      (1) To measure linearity, repeatability, half-value layer, and other data of the x-ray system.
      (2) To measure the intensity of the radiation exposure field at a given point to evaluate a potential radiation hazard.
      (3) To identify total exposure or rate of exposure.
   b. Test equipment:
      (1) *Autoranging digital dosimeter for diagnostic quality control:* Measures mA/time reciprocity, linearity, consistency/reproducibility, timer accuracy-tracking, and radiation output compared to standard unit. It also measures fluoroscopic exposure rates.
      (2) *Diagnostic x-ray calibrator:* This measures similar parameters as above.
      (3) *Direct reading dosimeters with chargers:* These dosimeters measure

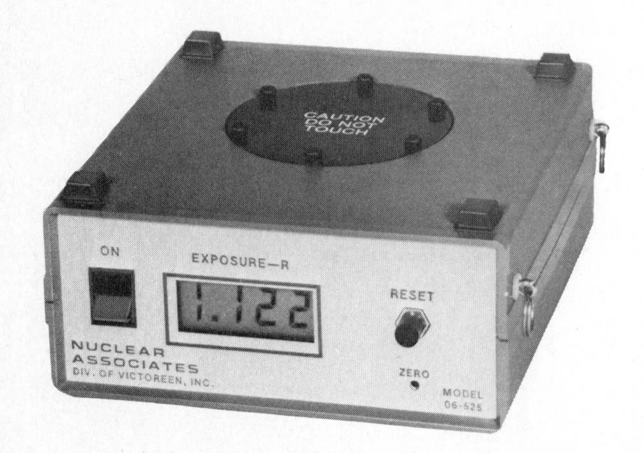

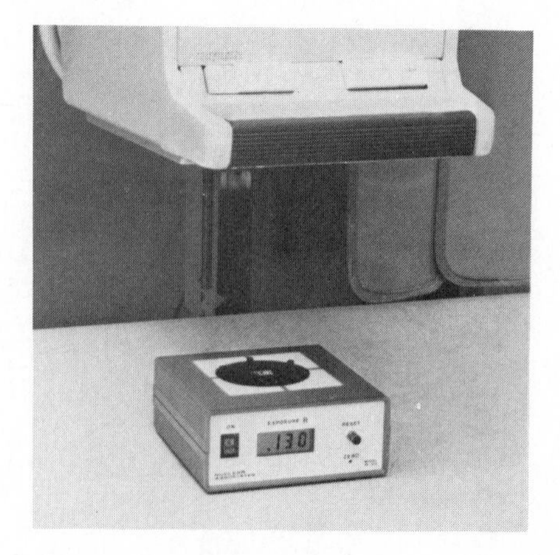

X-Ray QA Instrument
Courtesy of Nuclear Associates, Inc., Carle Place, N.Y.

implanted needles, x-ray systems, radioisotope generators, and operators of radiation therapy equipment. There are several types available.

c. Other test equipment:
   (1) Dosimeter kit for diagnostic x-ray quality control
   (2) Indirect reading dosimeter with charger
   (3) Low-energy radiation dosimeter with charger
   (4) Ion chamber
   (5) Strad Stray Radiation Dosimeter

d. Kits and multiple parameter devices:
   (1) X-ray Performance Kit
   (2) Dyanalyzer II
   (3) Basic Quality Assurance Test Kit
   (4) Others

5. *Focal spot measurements:* One of the factors limiting the resolution (sharpness) is the focal spot of the x-ray tube. The focal spot of the x-ray tube is *not* a point source. A blurred or fuzzy penumbra area degrades the x-ray image. This procedure is not a routine quality control monitoring parameter but is important in the evaluation of the deterioration of the focal spot. This parameter is usually checked during a preventive maintenance system check by the service engineer.

   a. Test equipment:
      (1) Delux Focal Spot Test Tool
      (2) Focal Spot Test Tool
      (3) Star Pattern Model Sp-300
      (4) Star Test Pattern
      (5) Wisconsin Focal Spot Test Tool (very specific)
      (6) X-ray Pinhole Camera
      (7) Others

   b. Kits and multiple parameter test devices:
      (1) Basic Quality Control Assurance Test Kit
      (2) X-ray Test Pattern

6. *Measurement of beam alignment, dimensions, and SID:* These parameters are measured to ensure that the x-ray radiation beam is exposing only the area of interest. It is the actual measurement of beam alignment and/or field size.

   a. Test equipment:
      (1) *Beam alignment and template:* This device checks x-ray light field versus radiation field.
      (2) *Beam size ruler:* Measurement of the beam size at the table top.
      (3) *Beam diameter gauge:* Measures the beam diameter of dental x-ray machine.
      (4) *17 Pin gauge:* Used to locate the central ray of the x-ray field.
      (5) *Wisconsin Beam Alignment Test Tool:* This device is designed to provide a simple pass-fail test for beam centering and perpendicularity to the image receptor at 1m (40 in.) SID.
      (6) *Wisconsin Collimator Test Tool:* Part of a radiologic quality control program is to measure the x-ray field and the x-ray field alignment. This device is adaptable to any x-ray system with a light field. Again it gives a simple pass-fail test at 1 m (40 in.) SID.

b. Kits and multiple parameter devices:
 (1) Basic Quality Assurance Test Kit
 (2) X-ray Performance Kit

7. *Film-screen monitoring:* Detection of faulty or damaged cassettes or dirty screens.
 a. Test Equipment:
 (1) *Cronex Test Grid:* This device is used to determine if there is poor film-screen congruence in the cassette.
 (2) *Scan-A-Screen:* An ultraviolet device is used to scan over the surface of intensifying screens to detect the presence of dust, dirt, and/or defects.
 (3) *Wisconsin Film-Screen Contact Test Tool:* This tool tests for film-screen congruence.

8. *Half-value layer:* This is a common methodology used in testing beam quality. Such an evaluation determines if the tube contains the amount of inherent filtration recommended by the ICRU or other regulatory bodies. It is also used as a parameter to assess tube deterioration.
 a. Test equipment:
 (1) *Diagnostic Half-Value Layer Kit:* Measurement and documentation at regular intervals of x-ray beam quality; device includes sample quality control recording forms.
 (2) *Half-Value Kit*
 b. Kits and multiple parameter devices:
 (1) Basic Quality Assurance Test Kit
 (2) Diagnostic X-ray Analyzer
 (3) Diarad I Dosimeter System

9. *Technique Calculators:* Aids in the variation of technique factors from standard values.
 a. Test equipment:
 (1) Radiographic Technique Conversion Guide
 (2) Supertech Pocket Technique Computer
 (3) Supertech Wall Mount Technique Computer with Rare Earth Converter
 (4) Tec X-ray Exposure Calculator

10. *Penetrometer and step wedge:* A device to check x-ray system that has various thicknesses yielding various radiographic densities.
 a. Test equipment:
 (1) *Multi-Mini-Wedge:* May be used with all tools that involve evaluating a series of densities.
 (2) *Patient Phantom/Penetrometer System:* This checks the table top radiation output of image intensified fluoroscopic equipment.
 (3) Step wedge
 (4) Others

11. *Miscellaneous supplies:*
 a. Test equipment:
 (1) *Calipers:* Measures the thickness of anatomical parts.
 (2) Quality control data forms
 (3) Measure Mate Patient Thickness Calculator
 (4) Targaret-To-Panel Gauge: Measures the targaret to panel table top distance.

III. SPECIALIZED X-RAY UNIT QUALITY ASSURANCE

A. Fluoroscopic Units

1. There are higher amounts of patient radiation dose due to longer exposure times.
2. *Test equipment:*
   a. *Fluoroscopic Beam Alignment Device:* This device is used to determine the alignment of x-ray field and image receptor.
   b. *Fluoroscopic System Resolution Test Tool:* This device is used to check the resolution of the panel and the image amplifier system.
   c. *Wisconsin Copper Mesh Tool:* This device is also used to check resolution.
   d. *Wisconsin Phantom and Penetrometer:* This device measures low contrast performance, automatic brightness stabilization, and output in roentgens under maximum conditions.
   e. Others
3. *Kits and multiple parameter devices:*
   a. Basic Quality Assurance Test Kit
   b. Diagnostic X-ray Analyzer
   c. Diarad I Dosimeter System
   d. Others

B. Tomographic Unit

Movement of the tomographic unit adds an additional dimension to its evaluation of performance. There have been some advances recently in the manufacturing of some test devices to deal with this parameter.

1. *Test equipment:*
   a. *Wisconsin Tomographic Test Tool:* Routine measurement of cut position and thickness and beam trajectory is a part of the quality control program. This equipment measures all these parameters.
2. *Kits and multiple parameter devices:*
   a. Basic Quality Assurance Program Test Kit

Hand-Held Dual-Color Sensitometer
Courtesy of Nuclear Associates, Inc., Carle Place, N.Y.

SENSITOMETERS

Type	*Information*	*Cost (1982)*
Du Pont Cronex Sensitometer	E.I. du Pont de Nemours & Co. Photo Products Department Wilmington, DE 19898	$ 675.00
Pro-Cal Sensitometer	Picker Corporation Corporate Headquarters 595 Miner Road Cleveland, OH 44143	270.00
3M X-ray Process Control Sensitometer	3M Company 3M Center X-ray Products Building 223 St. Paul, MN 55101	625.00
Wejex Model R Sensitometer, Developer Controller	Tobias Associates 50 Industrial Dr. Northampton Industrial Park Ivyland, PA 18974	225.00
Wejex Model RF, Sensitometer, Developer Controller	(Same as above)	525.00
Wisconsin Sensitometer Model 201	Radiation Measurement Inc. Box 44, 7617 Domar Dr. Middleton, WI 53562	145.00
X-Rite Model 303 Sensitometer	X-Rite Company 4101 Roger Chaffa Dr. S.E. Grand Rapids, MI 49508	545.00
Kodak Process Control Sensitometer Model 101	Eastman Kodak Co. Radiography Markets Div. Rochester, NY 14650	2,025.00
Gamma Check System-Flasher-Reader	Texas Medical Instruments Inc. 12108 Radium San Antonio, TX 78216	1,600.00 1,600.00
Quali-Trol II	Nuclear Associates, Inc. 100 Voice Road Carle Place, NY 11514	470.00

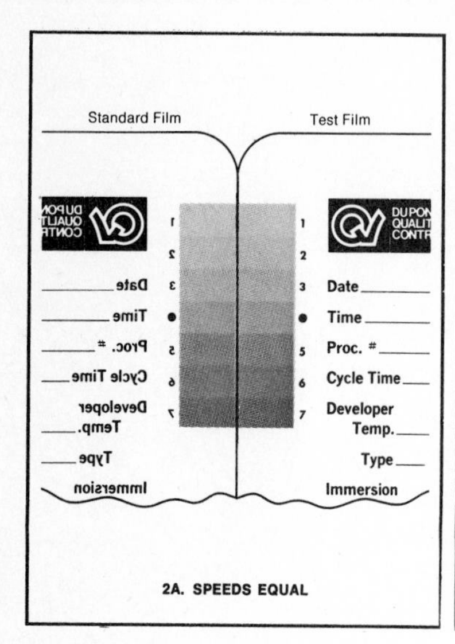

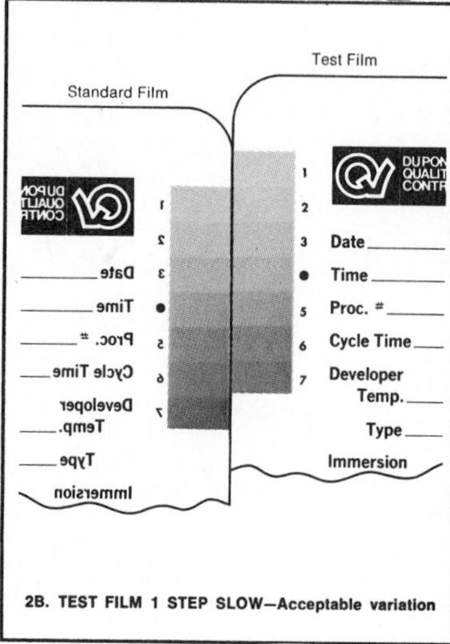

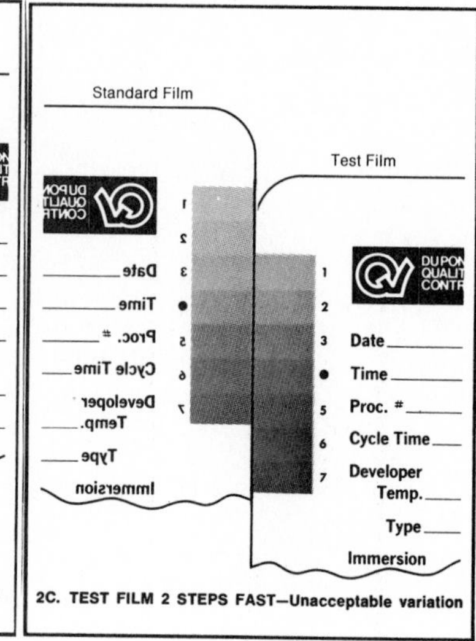

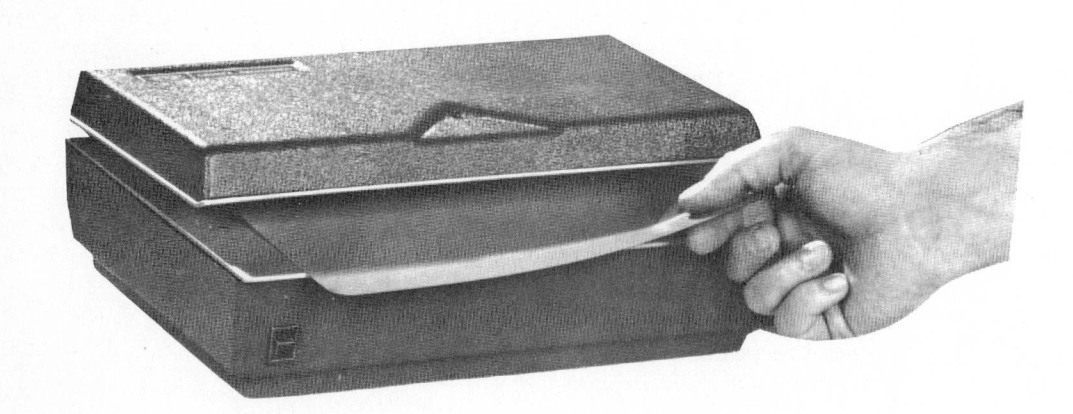

Sensitometer and Sensitometric Strips

Courtesy of E. I. duPont de Nemours & Co., Inc., Wilmington, Del.

SENSITOMETER CALIBRATION

Overview

One of the most difficult instruments to calibrate in the entire quality control program is the sensitometer. It is important to note that all measurements are *relative*. The radiographic or sensitometric exposures used in the quality assurance program need not be known on an absolutely precise basis.

TYPES OF SENSITOMETERS:

1. Intensity modulated sensitometer with a timing shutter:
   - a. Light source is always on.
   - *b.* The intensity is modulated by a sensitometric step tablet.
   - c. The user must maintain log of the total time the light source is on.
   - *d.* Keep extra bulbs on hand.
   - e. Instructions should be included on how to set ampere meter as well as the bulb location.
   - *f.* It is important not to let the amperage increase at any time beyond that set by the manufacturer.

2. Intensity-modulated sensitometer with electronically controlled exposure time:
   - a. Operates on capacitive discharge principle.
   - *b.* The intensity and the exposure time are controlled electronically.
   - c. The exposure time, in some models, can be precisely changed.
   - *d.* This sensitometer contains the sensitometric step tablet as a component.
   - e. No calibration is provided by the manufacturer for the determination of the actual exposure time.

Types 1 and 2 sensitometers most closely relate to or simulate the exposure received by a radiograph in a cassette:

3. Time-modulated sensitometer:
   - a. Varies exposure times.
   - *b.* Does not use step tablet.
   - c. Not recommended for radiographic quality control or quality assurance program.

GENERAL COMMENTS

1. If it is necessary to change the intensity of the sensitometer, a filter may be used.
2. In any type of sensitometer the overall light intensity may be reduced by the addition of neutral density filters.
3. Flashed density film may be substituted for neutral density filters. They may be obtained from a photographic supply house.

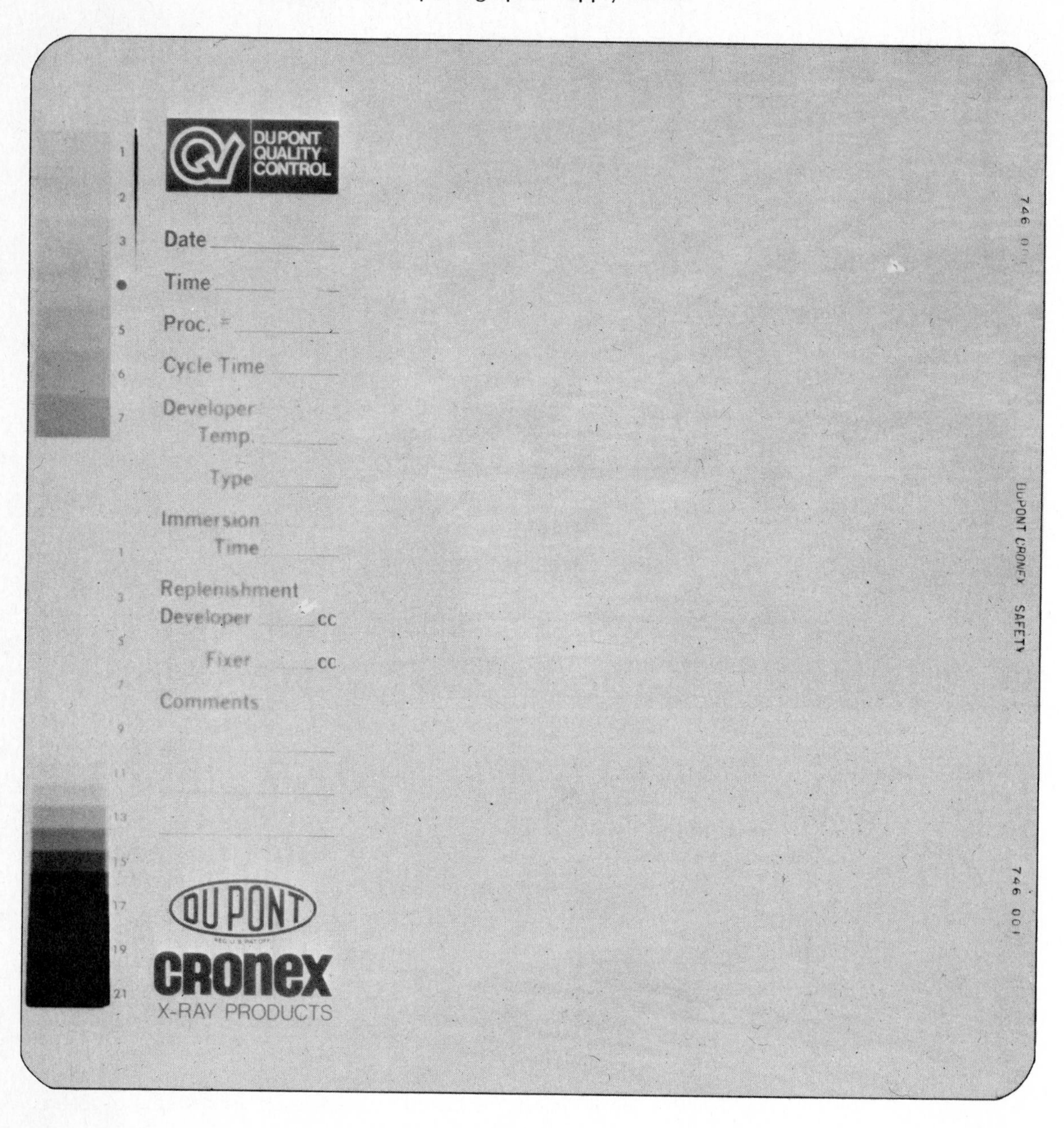

Sensitometric Strip
Courtesy of E.I. duPont de Nemours & Co. Inc.

SENSITOMETRIC USAGE LOG

Model # ______________________

Bulb # ______________________

Bulb Current ______________________

Date	Time On	Time Off	Cumulative Time	Initials

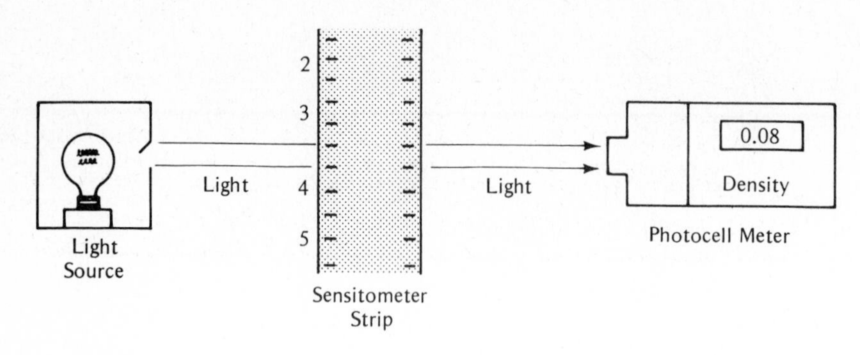

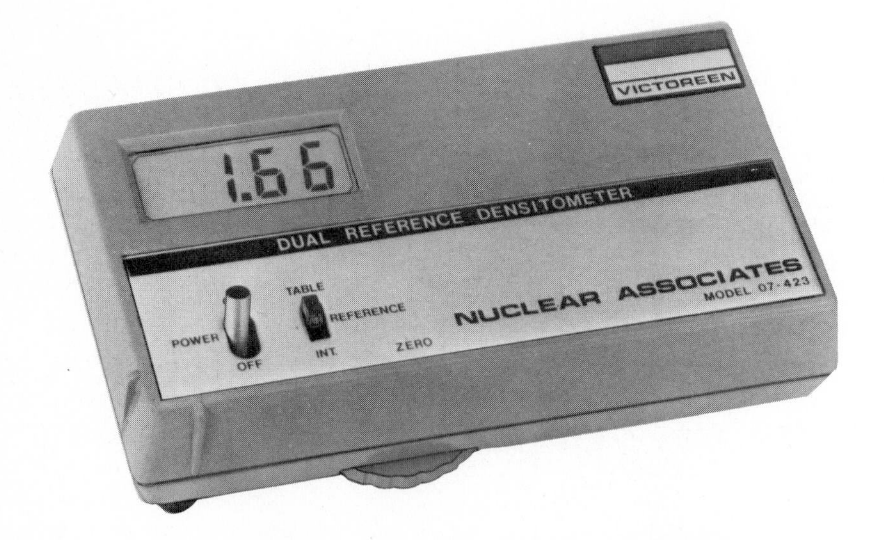

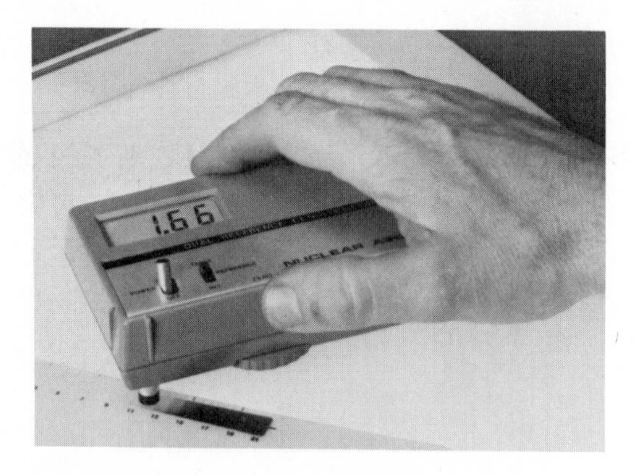

Dual-Reference Digital Densitometer

Courtesy of Nuclear Associates, Inc., Carle Place, N.Y.

DENSITOMETERS

Type	Information	Cost (1982)
Densitometer, Digital, Black and White Transmission, Model TBX	Tobias Associates, Inc. 50 Industrial Park P.O. Box 2699 Ivyland, PA 18974	$ 595.00
Densitometer Model HD-300 Photo Densitometer Hand Held	Telstar Electronics Corp. 700 Hummel Ave. Southold, NY 11971	260.00 260.00
Digital Densitometer	Nuclear Associates 100 Voice Road Carle Place, NY 11514	595.00
Macbeth Models 494TD500, 494TD102 494TD504	Independent x-ray dealers	850.00 1,175.00 1,575.00
Photovolt Densitometer	Independent X-ray Dealers Association 3142 Superior Ave. Cleveland, OH 44114	800.00
Sakura PDA-81	Sakura Medical 57 Bushes Land Elmwood Park, NJ 07407	265.00
3M X-ray Process Control Densitometer	3M Company 3M Center X-ray Products Building 223 St. Paul, MN 55101	649.00
Wisconsin Densitometer Model 211	Radiation Measurements Inc. Box 44–7617 Donna Dr. Middleton, WI 53562	245.00
X-Rite Model 301 Densitometer	The X-Rite Company 4101 Roger Chaffe Dr., S.E. Grand Rapids, MI 49508	795.00

DENSITOMETER CALIBRATION

Overview

When a new densitometer is received from a manufacturer, it will be calibrated. It is extremely important that you receive a calibration step tablet on which the calibrated density values for each step are documented from the manufacturer. This step tablet should be calibrated over a range of 3.0 density with a density difference between the steps of no more than 0.30.

COMPONENTS OF A DENSITOMETER

1. Light source
2. Sampling aperture
3. Optics
4. Photoelectric pickup
5. Some type of readout

WORDS OF WISDOM

1. If your original calibration step tablet is misplaced, lost, or damaged, a new one may be acquired from most professional photographic supply houses.
2. After receiving a new densitometer, verify that it is properly and accurately calibrated according to the operating instructions.
3. Precise results from the densitometer can only be obtained if it contains a voltage-regulated component part for the light source. External light sources such as view boxes allow for excessive variability.
4. Electronic densitometers are calibrated in density units of 0.02. Visual densitometers are calibrated in units of 0.05.
5. Always measure the density in the middle of the step wedge. The coefficient of error between the wedge values and those of the densitometer should not exceed ±0.02 to 0.03, depending upon the type of densitometer.

DENSITOMETER CONTROL CHART

INSTRUMENT: ______ MODEL: ______ SERIAL # ______

CALIBRATION: ______ STANDARD WEDGE # ______

+0.08

______

High

−0.08

+0.08

______

Med.

−0.08

+0.08

______

Low

−0.08

Initials

Date

SYSTEMS GEOMETRY

Overview

Systems geometry is made up of four important areas. They are as follows:

1. Focal spot size
2. Filtration-Collimation
3. Systems alignment
4. Patient-image intensifier tube relationship

Focal Spot Size

Effects of effective focal spot (what you actually see)

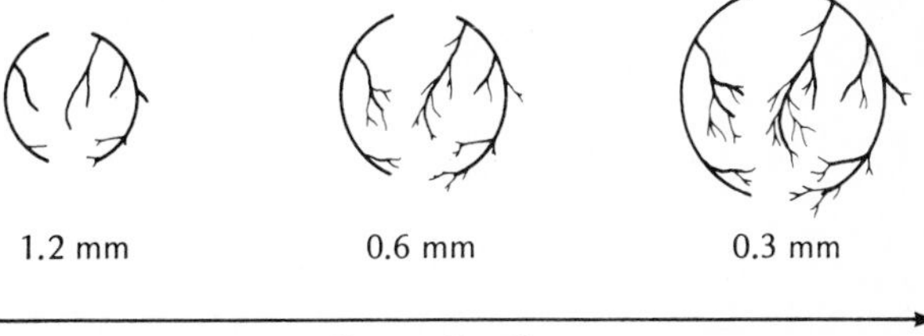

1. The x-ray tube is usually the limiting factor. However, there have been great strides in this area in the past few years, for example, x-ray tubes with heat factors of over 1,000,000 heats units.
2. Need 1 degree for each inch covered for angulation of anode.
3. One must know what size the input phosphor is. The greater the angle, the more one is able to load it.
4. The smaller the focal spot, the greater the resolution and detail.
5. The image tube should be as close to the patient as possible without hitting him.

X-ray Beam Filtration

1. Location of filter
2. Collimation 2–3 stage. Cuts down on increased kV penetration.
3. Quality of kV beam:
   a. Penetration: 85% primary beam absorbed by the patient
   b. Scatter
   c. Contrast

   As kV increases, scatter and penetration increase and contrast decreases.

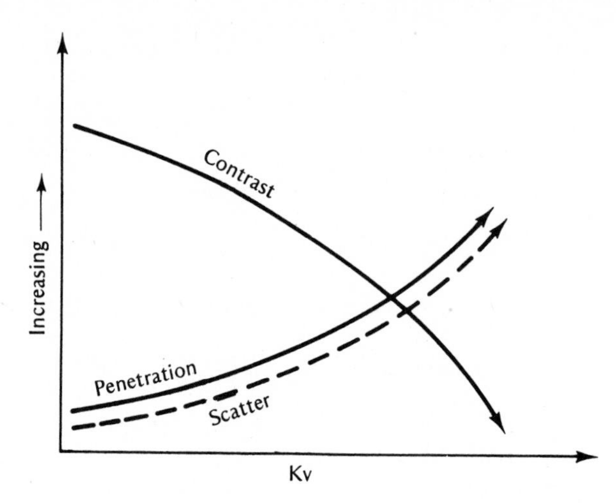

System Alignment

1. 90 degree alignment from x-ray tube → image intensifier tube → camera.

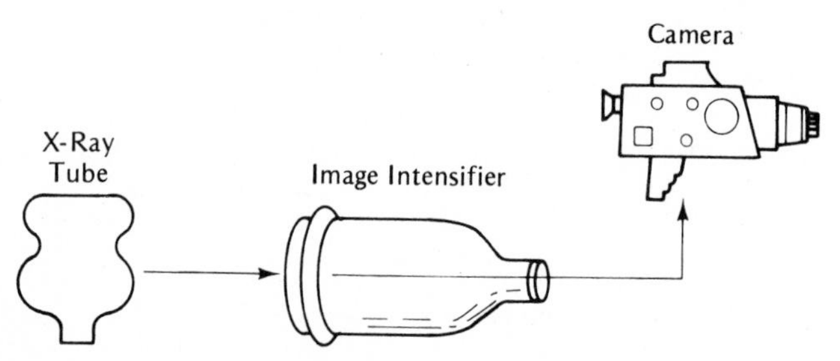

2. *Test alignment:* Place a piece of pipe on the table. Move the image intensifier back and forth noting any distortion.

Patient-Image Intensifier Relationship

Referring to the previous diagram, place the patient between the x-ray tube and the image intensifier tube. The x-ray tube, patient, image intensifier system, and the camera all need to be in alignment.

THREE SYSTEMS GEOMETRY TOPICS

1. Automatic dose control: phototiming
2. Pulse width
3. Milliamperes

Phototiming

1. Photo cell: attached to the image intensifier.
2. Automatic synchronization: camera shutter is in control of the system.

Pulse Width

1. Pulse width is time.
2. The time wave is a square wave.
3. Increased pulse width, decrease in quantum mottle.
4. Energy from the x-ray tube increases with time.
5. Amount of light to film from image intensifier increases with time.

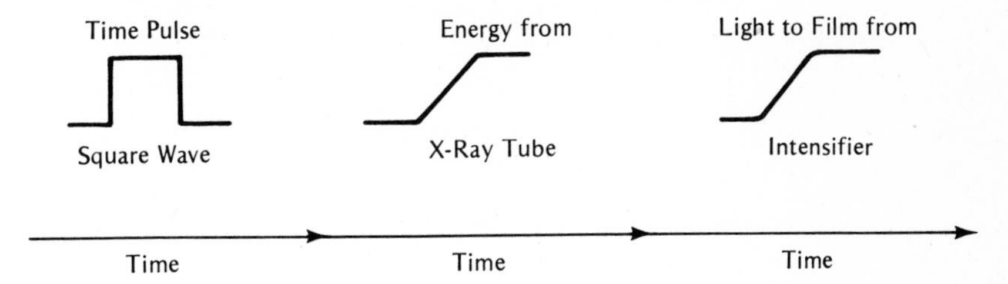

Milliamperes

1. Milliamperes (mA) is quantity.
2. Increase in mA, decrease in quantum mottle, increase in the radiation dose.

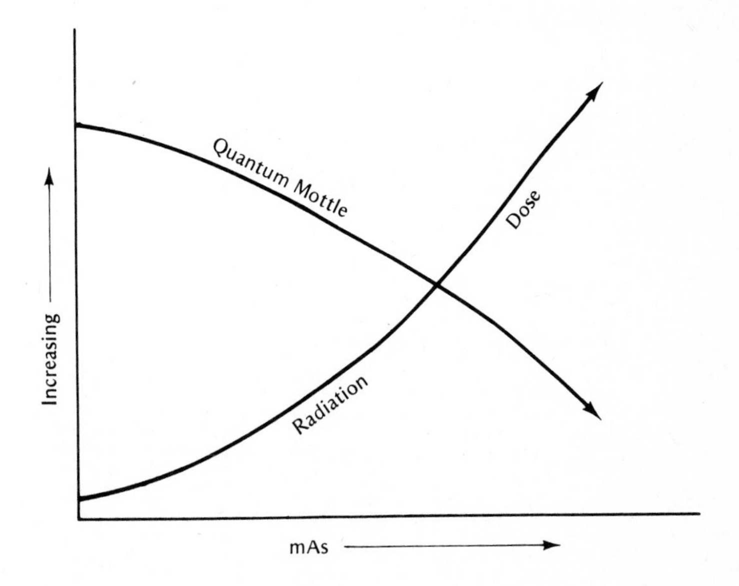

IMAGE INTENSIFIER: HOW DOES IT WORK?

1. Image amplification:
   a. Minification of the image
   b. Electronic acceleration
2. What does it look like?

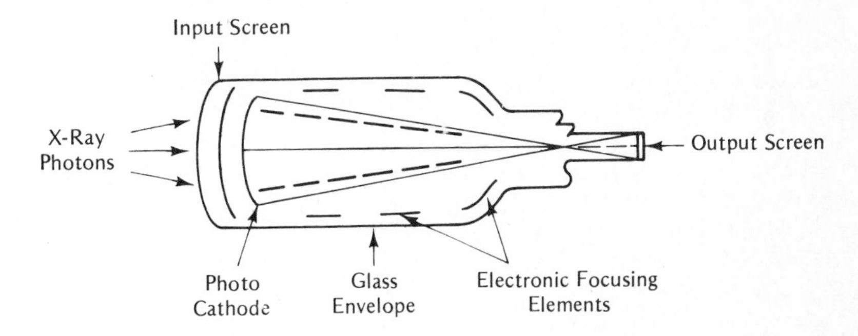

X-ray beam —▸ input screen —▸ light to photocathode —▸ electron flow (flow directed by electronic focusing elements) —▸ output phosphor screen.

Diagrammatically,

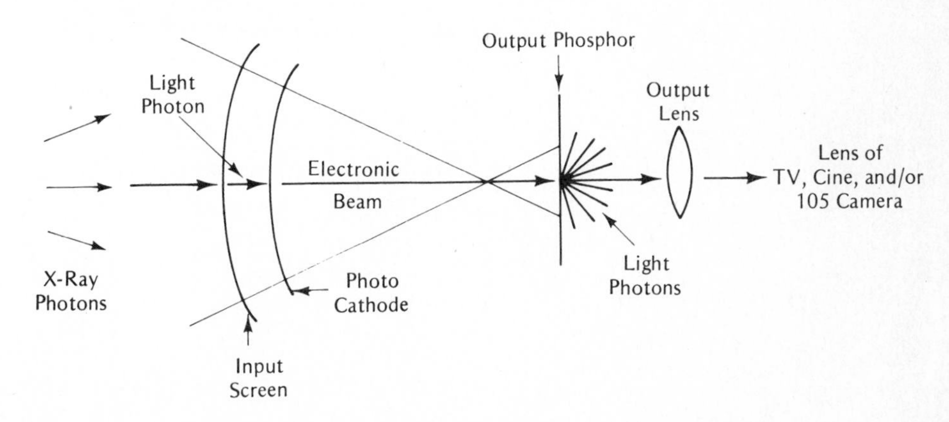

PHYSICAL IMAGE FUNCTIONS

1. *Density:* the inherent blackening of the film.
2. *Maximum density:* the blackest portion of the film.
3. *Minimum density:* the clearest portion of the film. It is known as *base plus fog.*
4. *Contrast:* the difference between the minimum and the maximum densities.

Graphs Depicting Large Area Contrast

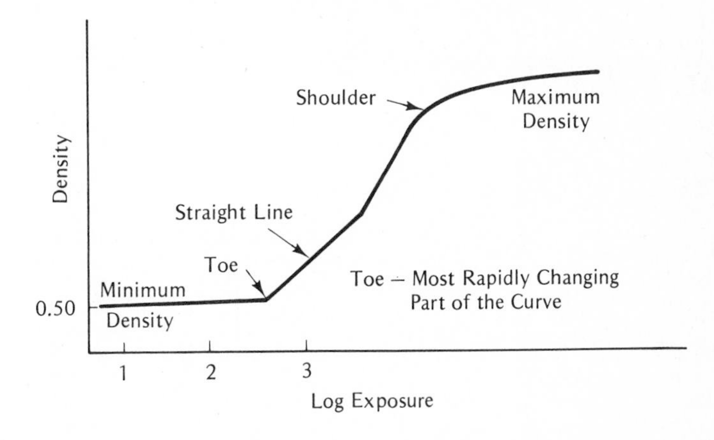

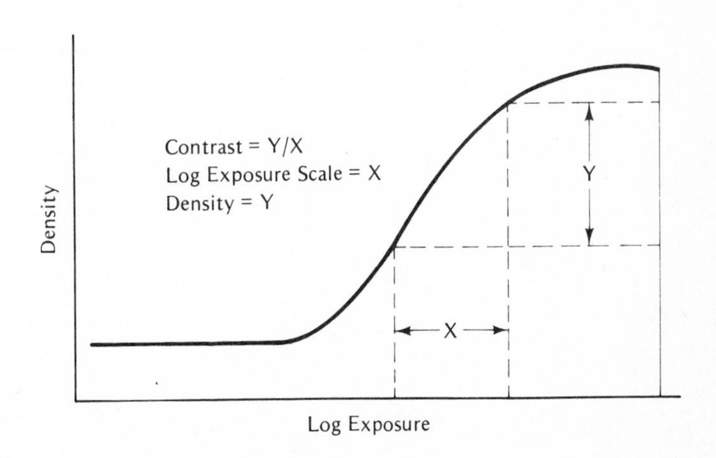

THREE FACTORS THAT INFLUENCE CONTRAST CURVES

1. Single person
2. Single process
3. Single processor

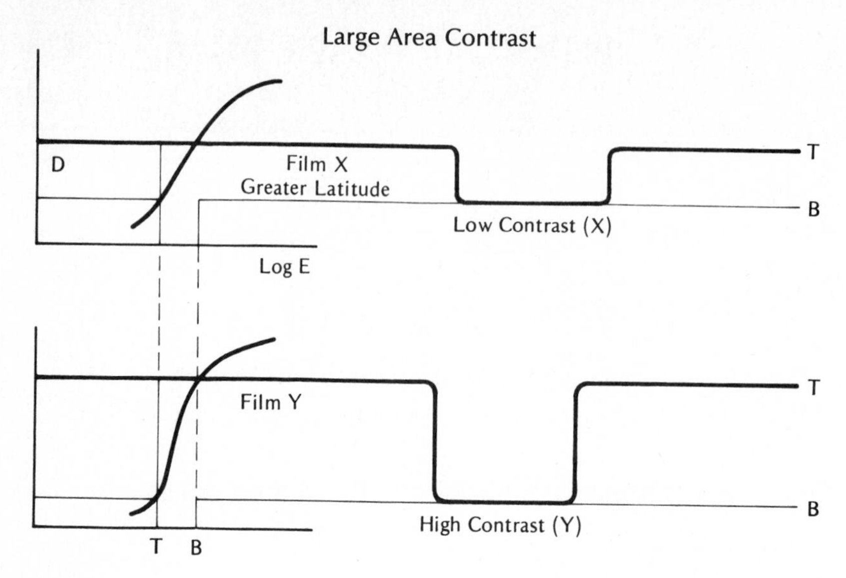

T = Tissue
B = Bone
D = Density
Log E = Log Exposure

Detail Contrast

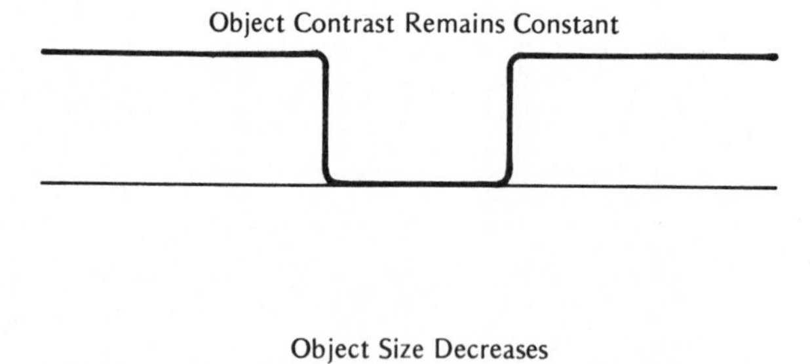

Object Size Decreases

Overall Contrast is Limited
No Reduction of Overall Subject Contrast

PHYSICAL IMAGE QUALITY EVALUATION

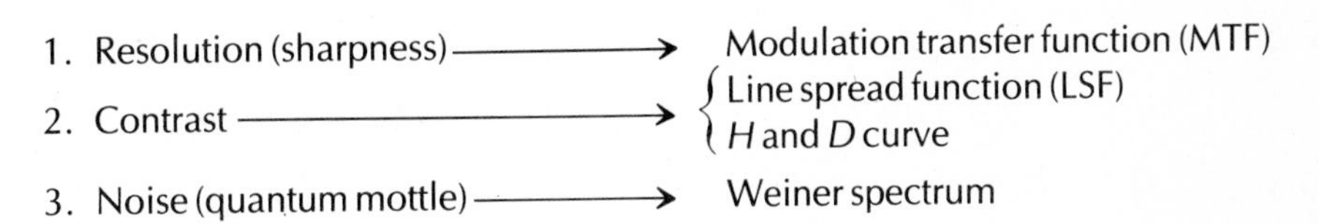

There is a direct relationship between the resolution, contrast, and noise and the physical image of the radiograph.

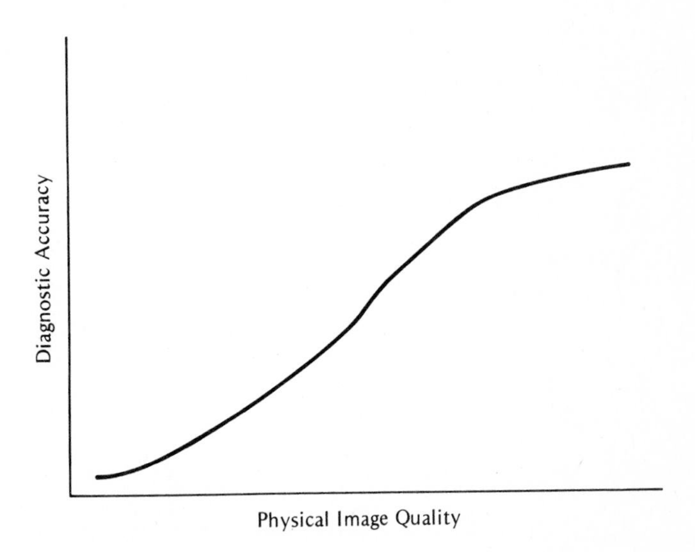

Diagram denoting the relationship between diagnostic and the physical image quality of the radiograph.

NOISE CONTRAST (QUANTUM MOTTLE)

Overview

Noise contrast (quantum mottle) is directly dependent upon the large area contrast, detail contrast, and the number of quanta used in the production of the image. The following figure is a line drawing of low and high noise systems.

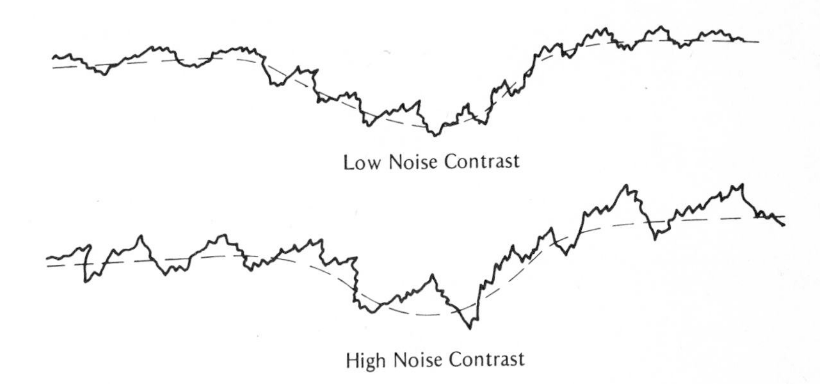

Problem

High level of noise on radiograph.

Possible solutions	*Rationale*
1. Decrease the speed of the film.	1. The greater the speed of the film, the fewer quanta used, resulting in a high noise level.
2. Utilization of a lower brightness screen of a similar phosphor composition.	2. A screen whose intrinsic brightness is high uses less quanta, resulting in a high noise level.
3. Use a film with less contrast.	3. Film contrast increases film radiation noise.

VARIABILITY IN RADIOGRAPHIC IMAGE RECORDING

Overview

There are many sources of variability resulting in a radiograph that is of less than optimum image quality. It must be noted that technologist error can play a role in poor image quality. Only considerations that may produce variations in exposure and that contribute to variations in the speed of the image receptor will be examined.

VARIABILITY IN THE RADIOGRAPHIC IMAGE RECORDING PROCESS

X-ray System

There are many numerous variables in this category. A few of these include radiation output, timing, condition of the anode of the x-ray tube, line voltage fluctuations, voltage waveforms, preheat settings, and regulation of the voltage source.

Radiographic Table-Grid

Inherent in all x-ray equipment is manufacturers' variation, which influences primary and scatter radiation.

Automatic Timing Devices (Phototiming)

Variations from the source may influence feedback control variations, sampling strategies, disparity between phototimer and image receptor kV response.

Patient–Technologist

Cooperation of the patient and expertise of the technologist are important factors.

Radiographic Processing

This is the crux of the quality control program and its problems. The variables are numerous. A few are temperature of the developer, water, and dryer, replenishment rate error (too high, too low), improper chemical mixing, improper transport speed, inadequate or improper preventive and corrective maintenance.

View Boxes, Hot Lamps, Projectors

There is not a great deal of variability in this category. Factors of light versus time, surface conditions of the items, and light intensity play a role in how the radiograph is viewed by the radiologists.

VARIABILITY OF IMAGE RECEPTOR ELEMENTS

Cassettes

The actual structural makeup of the cassette will influence absorption.

Films

Variability in this category will show in emulsion bases, sensitizing, and coating.

Screens

An influential factor of radiographic screens is the composition of the phosphor. Other factors are the amount of phosphor, substrate reflectivity, overcoat, and binder.

Radiographic Processor

The frequency at which this topic reappears tells us of its importance to the whole quality control process. As stated earlier, there are numerous variables in radiographic processing. A few are temperature of the developer solution, water, and dryer, transport speed, replenishment rate, recirculation time of the chemicals, and solution levels.

Radiographic Processor Chemicals

Only chemical purity and formulation variations are found in this category.

GRIDS

Overview

A grid is made up of alternate strips of lead and radio-opaque spacers of plastic or aluminum. As the x-ray beam penetrates a part of the body, the x-ray beam will penetrate the spacers but will not penetrate the lead strips. This results in less scatter and greater film contrast. X-rays do not penetrate the lead strips due to the density of the metal.

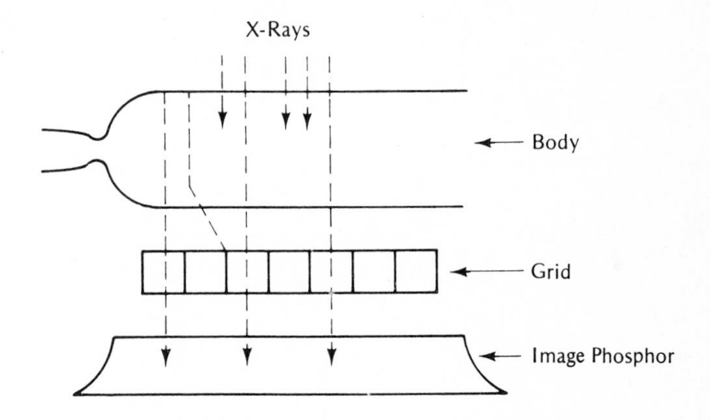

GRID RATIOS

Grid ratios are calculated by the height of the strip times the area between the strips.

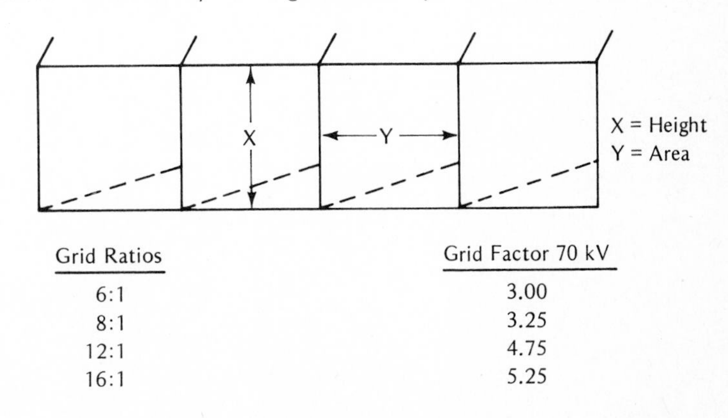

Grid Ratios	Grid Factor 70 kV
6:1	3.00
8:1	3.25
12:1	4.75
16:1	5.25

Facts on Grids

1. The higher the grid ratio, the greater the dose.
2. The higher the grid ratio, the greater the contrast.
3. The grid ratio affects the amount of radiation that reaches the input phosphor screen. This in turn yields a higher radiographic film contrast.

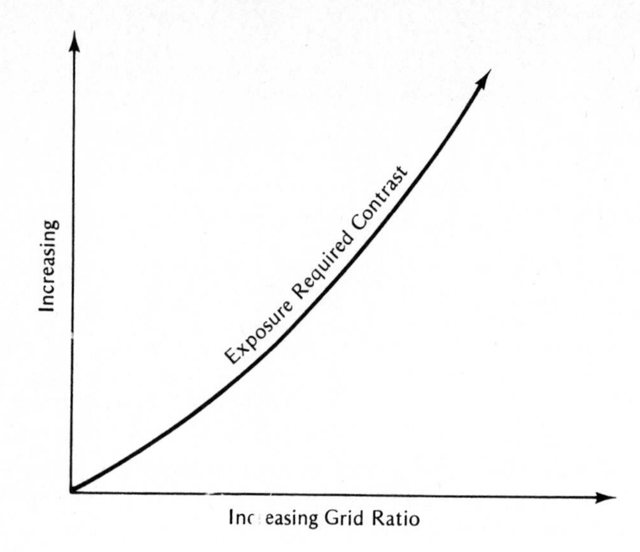

PROPERTIES OF RADIOGRAPHIC FILM

1. Optimum physical image quality
   a. Vascular clarity
   b. Resolution: sharpness
   c. Quantum mottle: noise factor
   d. Penetration:
      (1) Overexposed
      (2) Underexposed
   e. Contrast
2. Radiographic film terminology
   a. Color sensitivity: yellow-green most sensitive
   b. Resolution: sharpness
   c. Quantum mottle: noise factor, graininess (low to heavy)
   d. Density: blackness of the film
   e. Maximum density: darkest portion of the film
   f. Minimum density: clearest portion of the film (base plus fog)
   g. Gray base
   h. Straight line contrast: part of the H and D curve
   i. Time/Gamma
   j. Modulation transfer function (MTF)
   k. Characteristic curve: H and D curve
3. General construction of cine film
   a. Surface
   b. Emulsion: silver halide salt
   c. Base: gray dye particles
   d. Antistatic back coating
4. General construction of radiographic film
   a. Surface
   b. Emulsion: silver salts
   c. Base
   d. Emulsion: silver salts
   e. Surface

5. Film speed versus grain size
   a. The faster the film, the larger the grain.
   b. The slower the film, the finer the grain.

Construction of Radiographic Film

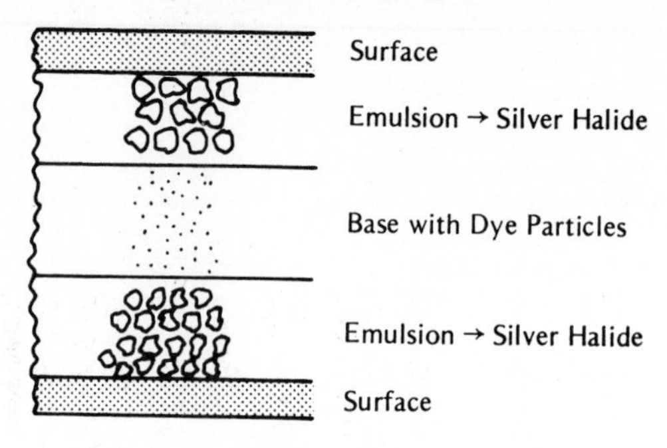

Construction of Cine Film

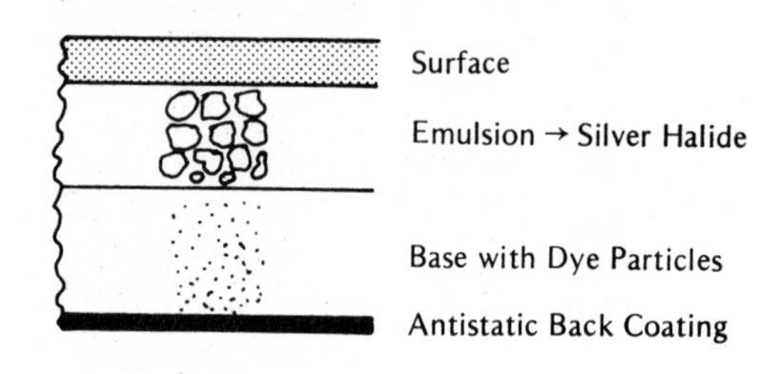

Film Grain

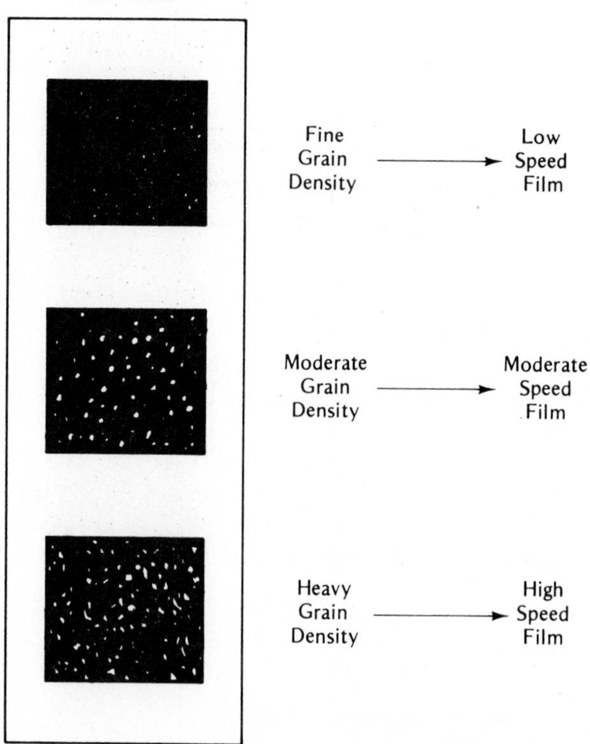

Film Grain: In cine and general radiography, the size of the film grain correlates with the speed of the film.

6. Modulation transfer function (MTF)
   - *a.* Calculation of line pairs per millimeter
   - *b.* Objective quantitative measure
   - *c.* Contrast versus film detail
   - *d.* Overall system MTF is obtainable:
     - *(1)* Image intensifier MTF
     - *(2)* Lens MTF
     - *(3)* Film MTF

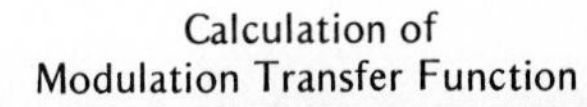

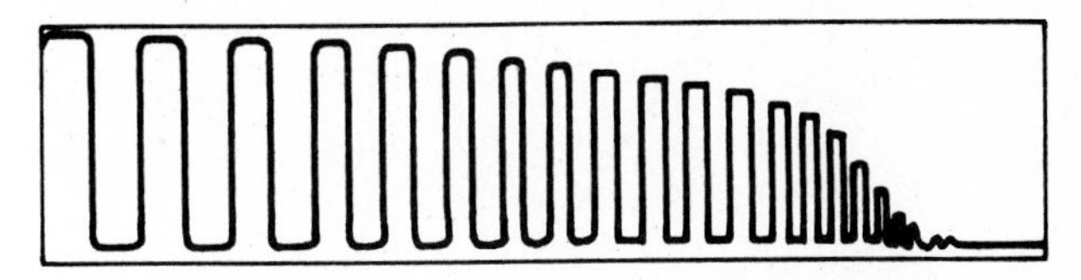

Microdensitometer Trace of Film Image of Target

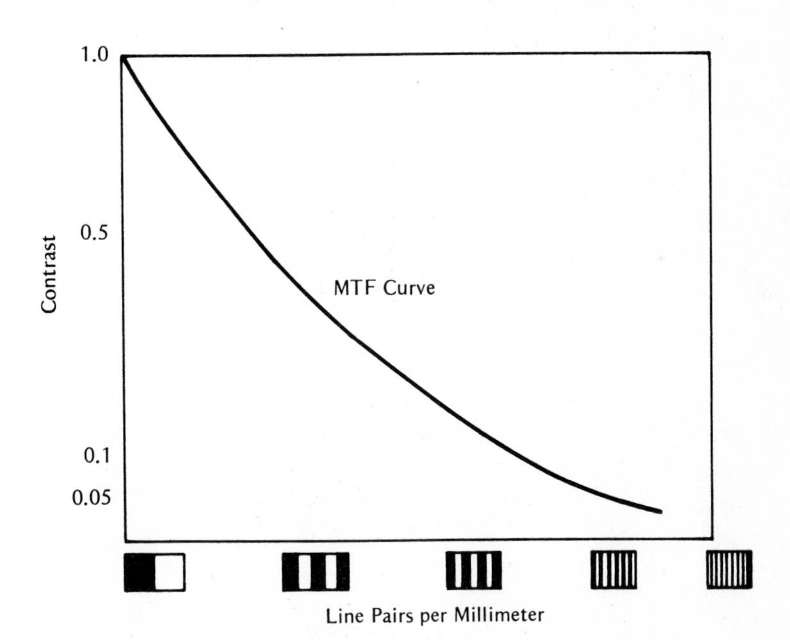

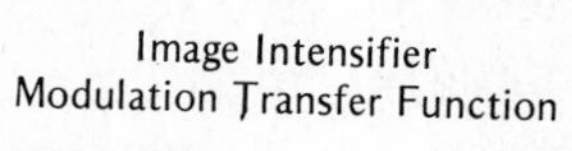
Image Intensifier
Modulation Transfer Function

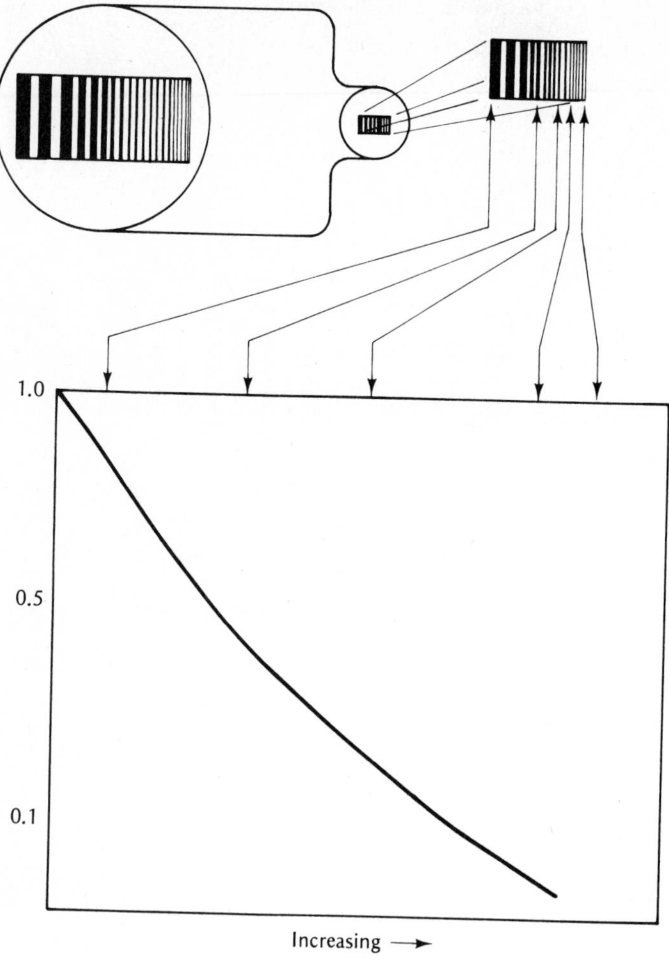
1.0
0.5
0.1
Increasing
Line Pairs per Millimeter

System MTF

Individual MTF's may be Multiplied Point by
Point to Obtain System MTF

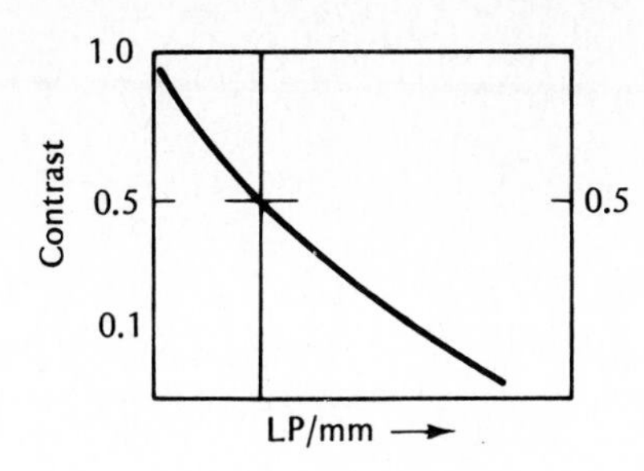

Image Intensifier MTF

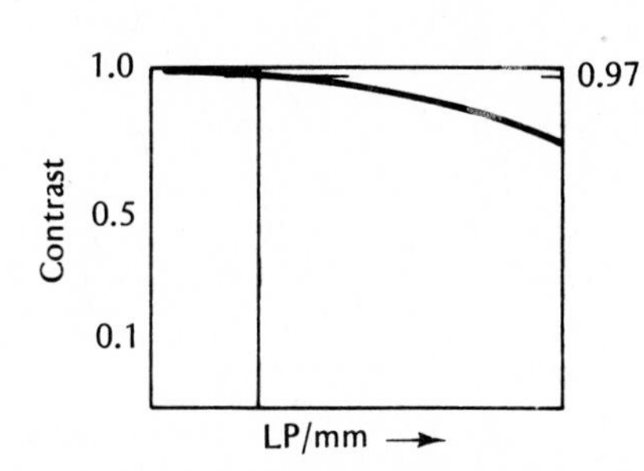

Film MTF

0.5 X 0.97 = 0.485

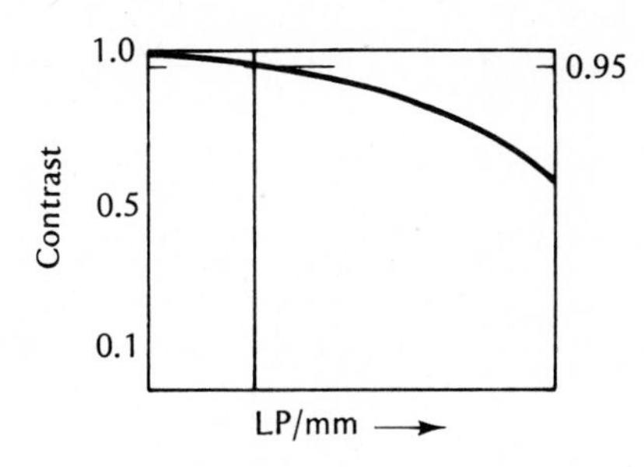

Lens MTF

0.485 X 0.97 = 0.46

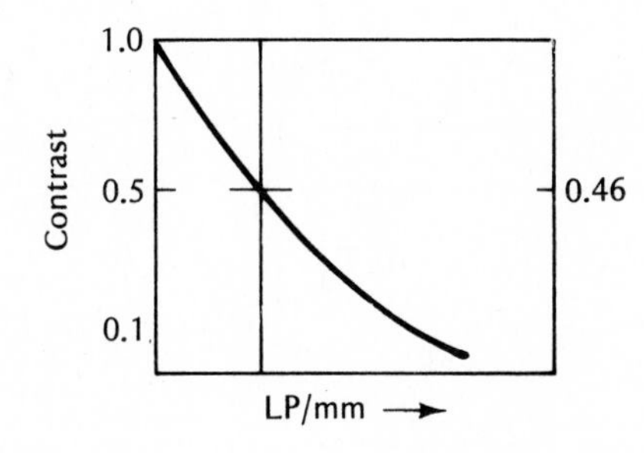

System MTF

LP/mm = Line Pairs per Millimeter

FACTORS THAT AFFECT FILM-SCREEN RESOLUTION

1. Contrast
2. Focal spot size and distribution
3. Magnification
4. Motion blurring
5. Noise
6. Radiographic processing
7. Screens

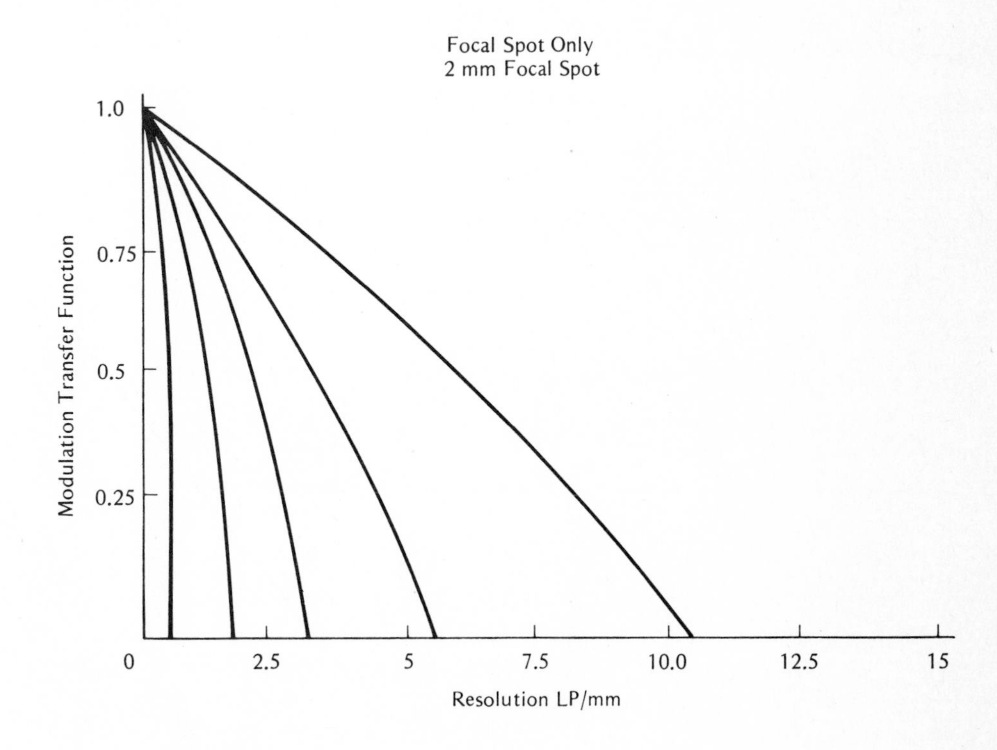

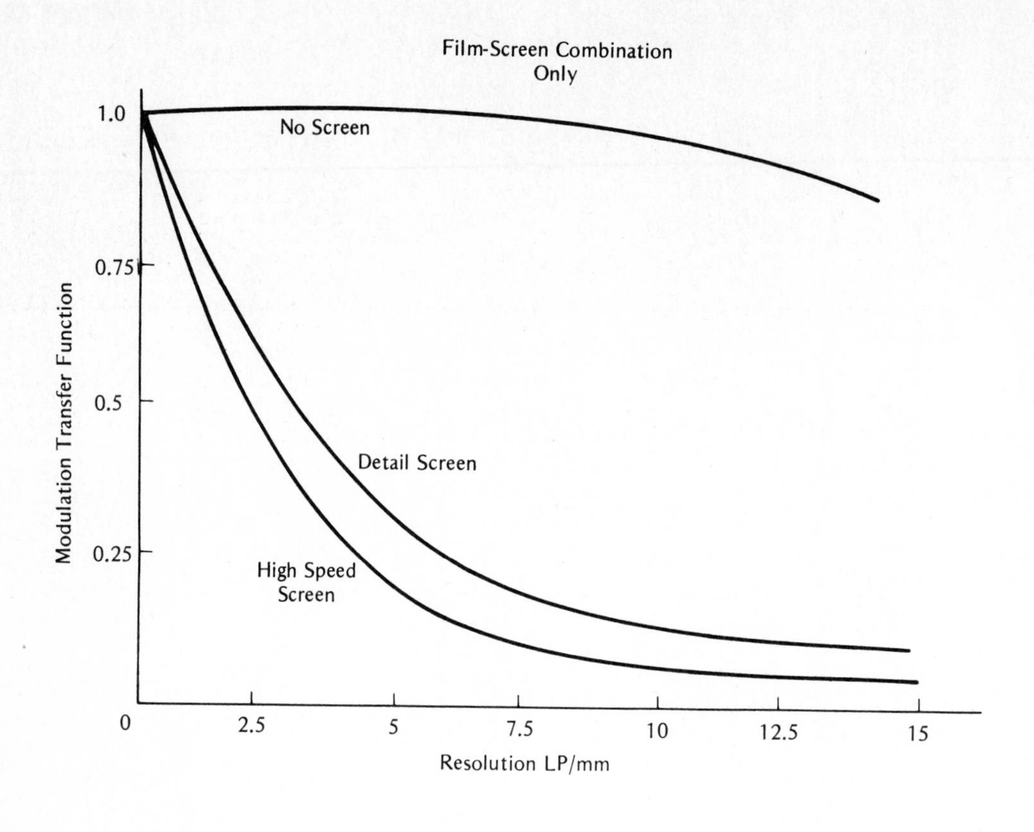
Film-Screen Combination
Only
1.0
0.75
0.5
0.25
Modulation Transfer Function
No Screen
Detail Screen
High Speed
Screen
0
2.5
5
7.5
10
12.5
15
Resolution LP/mm

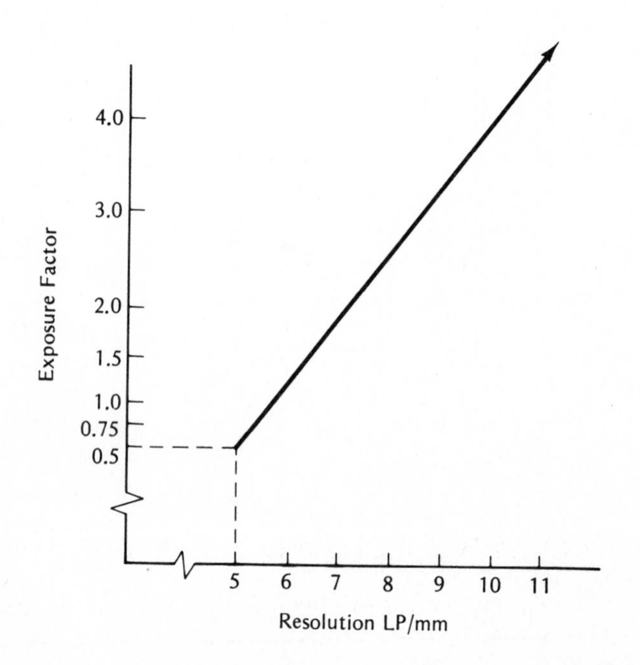
4.0
3.0
2.0
1.5
1.0
0.75
0.5
Exposure Factor
5
6
7
8
9
10
11
Resolution LP/mm

Screens

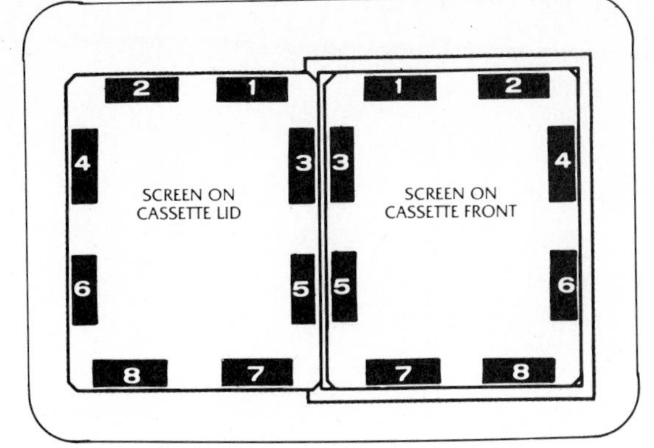

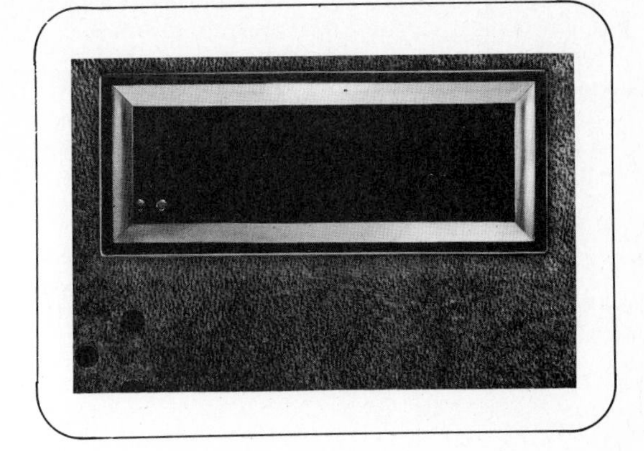

Cassettes

CRONEX MEDICAL X-RAY FILMS	CRONEX BLUE-EMITTING INTENSIFYING SCREENS			
	NEW Quanta III	Quanta II	Hi-Plus	Par
Cronex 4	Speed **800** **Cronex 4/Quanta III** Ultra speed, image similar to Cronex 4/Quanta II with minimal increase in noise	Speed **400** **Cronex 4/Quanta II** High speed, image similar to Cronex 4/Hi-Plus	Speed **200** **Cronex 4/Hi-Plus** Industry standard at medium speed, high contrast.	Speed **100** **Cronex 4/Par** Industry standard at speed, high contrast, best sharpness.
Cronex 2DC	Speed **800** **Cronex 2DC/ Quanta III** Ultra speed, high contrast, image clarity of Cronex 2DC.	Speed **400** **Cronex 2DC/ Quanta II** High speed, high contrast, image clarity of Cronex 2DC.	Speed **200** **Cronex 2DC/Hi-Plus** Medium speed, high contrast. Industry standard for 3½ minute processing.	Speed **100** **Cronex 2DC/Par** Industry standard at Par speed and 3½ minute processing, best sharpness.
Cronex 6 Plus	Speed **800** **Cronex 6 Plus/ Quanta III** Ultra speed, excellent low density contrast plus tissue visibility.	Speed **400** **Cronex Plus/ Quanta II** High speed, excellent low density contrast plus tissue visibility.	Speed **200** **Cronex 6 Plus/ Hi-Plus** Medium speed, excellent low density contrast plus tissue visibility.	Speed **100** **Cronex 6 Plus/Par** Par speed, excellent low density contrast plus tissue visibility, best sharpness.
Cronex 6	Speed **800** **Cronex 6/Quanta III** Ultra speed, wide latitude, medium contrast.	Speed **400** **Cronex 6/Quanta II** High speed, wide latitude, medium contrast.	Speed **200** **Cronex 6/Hi-Plus** Medium speed, wide latitude, medium contrast.	Speed **100** **Cronex 6/Par** Par speed, wide latitude, medium contrast, best sharpness.
Cronex 7	Speed **400** **Cronex 7/Quanta III** High speed, less noise than Cronex 4/ Quanta II, image clarity like Cronex 2DC.	Speed **200** **Cronex 7/Quanta II** Medium speed, high contrast, image clarity like Cronex 2DC.	Speed **100** **Cronex 7/Hi-Plus** Par speed, high contrast, lowest noise, image clarity like Cronex 2DC.	

Use CRONEX Cassettes with 19 film/screen combinations from Du Pont.

Courtesy of E. I. duPont de Nemours and Company, Inc., Wilmington, Del.

Exposure Requirements

Physical Image Quality \ Screens	Rare Earth Screens	Calcium Tungstate Screens
Medium Speed Screens	0.5 mR	1.0 mR
Fast Speed Screens	0.25 mR	0.5 mR

DIAGRAMMATIC PRINCIPLES OF FILM-SCREEN CONGRUENCE

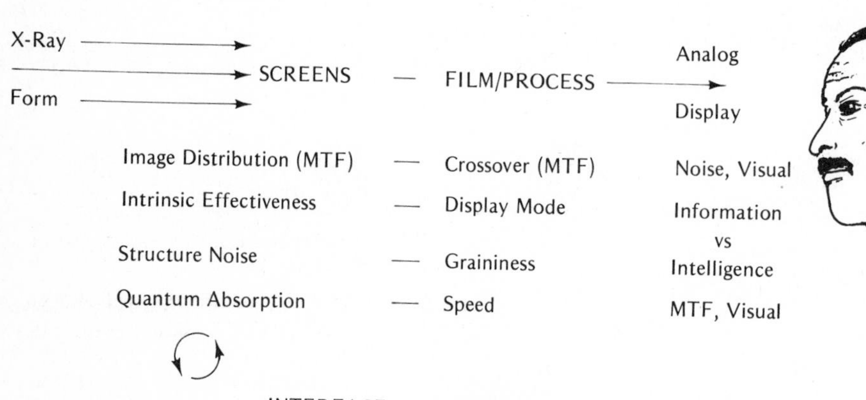

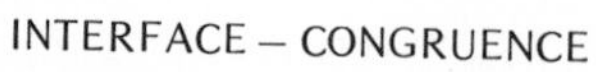

X-Ray Form	Emitted Radiation	Film Process	Emulsion
Screens	Radiation to Light	Analog Display	Film, T.V. Screen, Video, Disc., Etc.

CALIBRATION OF THE X-RAY SYSTEM

Overview

An important step in radiographic quality control is the proper calibration of the x-ray system. Without the proper calibration of the system, there could be no consistency in the radiographic film quality. The x-ray technologists could not utilize technique charts without proper calibration.

It is through optimum calibration of the x-ray system that similar techniques may be established in different rooms of radiographic equipment. These rooms need not be of the same type or brand of equipment for similar techniques.

There are many methods by which x-ray systems can be calibrated. One method is by utilizing an aluminum step wedge and spinning top. Another method is the utilization of a phantom. Both of these methodologies may be used for checking daily quality control by the quality control coordinator of the radiology department. In-house and/or contract service engineers have complex equipment that allows them to directly measure outputs and calibrate the x-ray system.

OBJECTIVE

To track the generator by use of an aluminum step wedge and/or phantom for radiographic quality control.

PLAN: GATHER EQUIPMENT

1. Aluminum 7 to 21 step wedge
2. Phantom
3. Control chart
4. Pencil

Implementation	*Rationale and key points*
1. Set a kVp and an mAs of the same value to obtain equal radiographic results. (See chart for an example.)	*1.1* Constant mAs: track mA and check the kVp compensation by changing time and mA for each of the mA stations.

Implementation	*Rationale and key points*
2. Use the aluminum step wedge or phantom to track the mA of the generator.	2.1. The resulting exposures obtained in using the step wedge or the phantom should be exactly the same density.

Responsibility	*Action*
1. Quality control coordinator	1.1. Assess the calibration of the x-ray system or systems on a daily basis. 1.2. Document the results of these quality control tests.

APPROVED: ____________________

DATE: ____________________

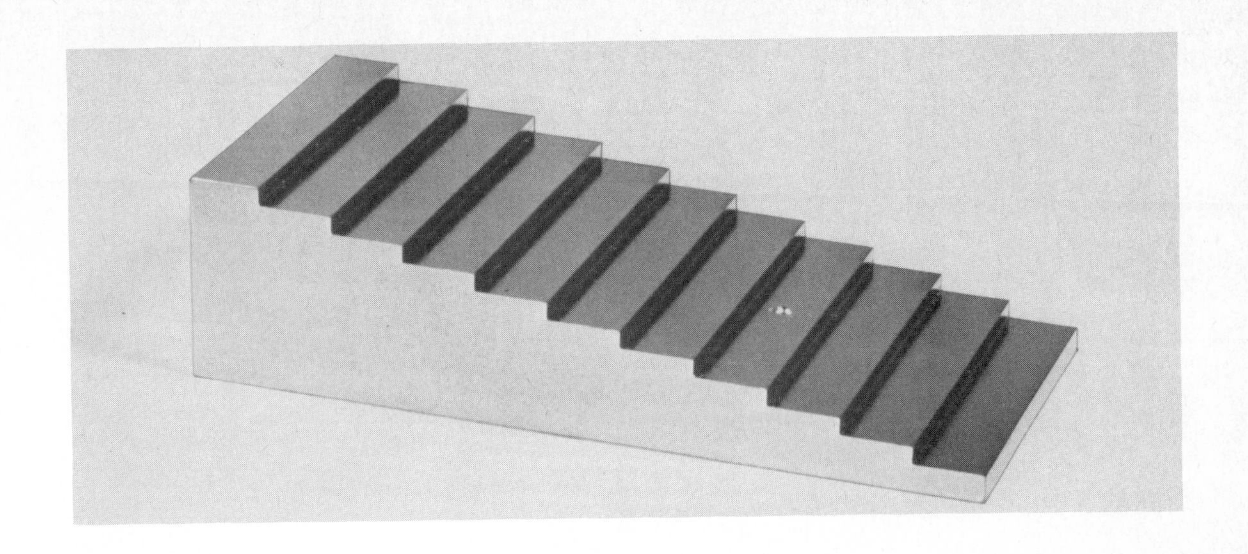

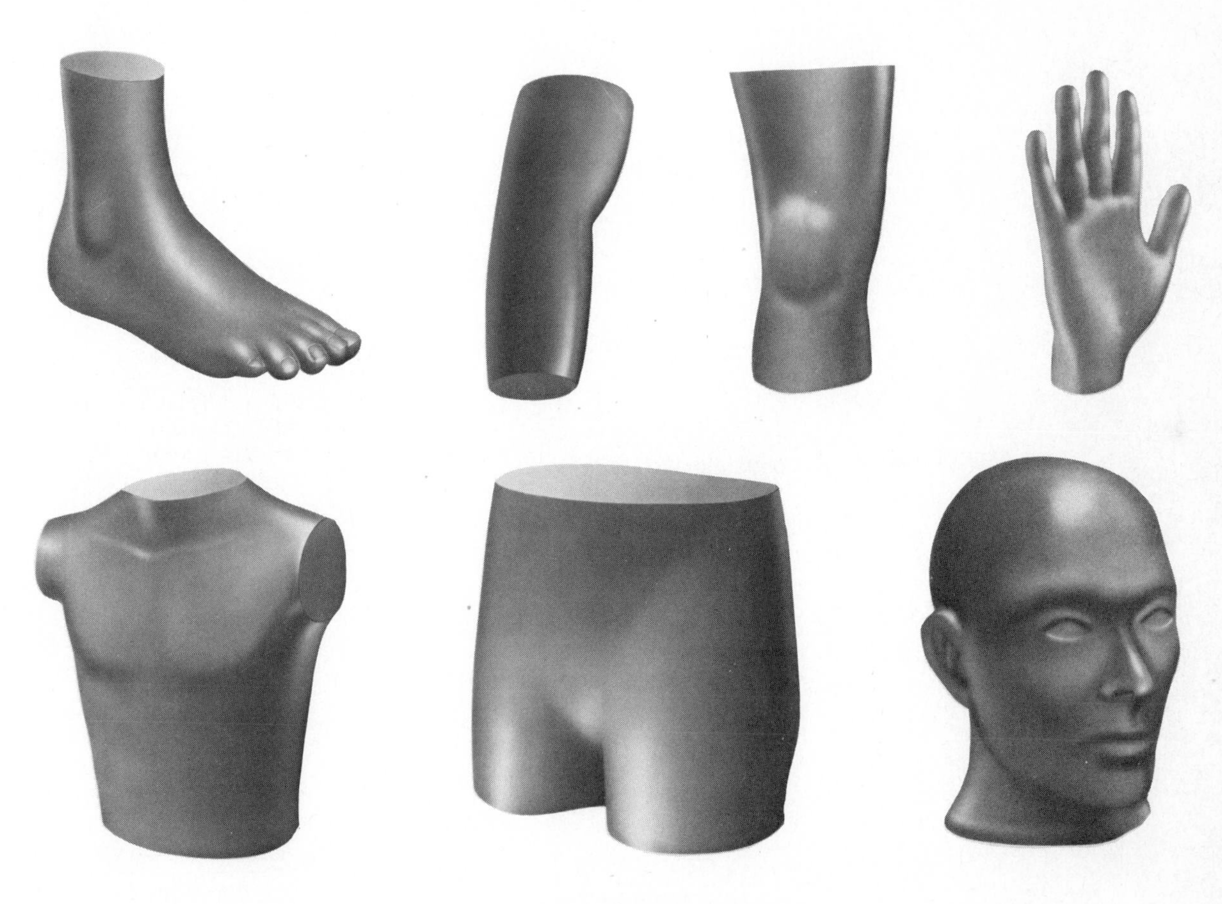

X-Ray Phantoms

Courtesy of Nuclear Associates, Inc., Carle Place, N.Y.

EXAMPLE OF CALIBRATION CHART

Use of 7 to 21 step wedge to track generator

	48-inch focal film distance (FFP) Table top
Known kVp and constant mAs	80 kVp 10 mAs

mA Stations	50 mA	100 mA	200 mA	300 mA
Time	'/5	'/10	'/20	'/30
1.				
2.				
3.				
4.				
5.				
6.				
7.				
8.				
9.				
10.				

USE OF A SKULL PHANTOM TO TRACK GENERATOR

Lateral skull	40-inch film focal distance (FFD) or source to image distance (SID) 12:1 Grid
Known kVp and constant mAs	80 kVp 40 mAs

mA Stations	50 mA	100 mA	200 mA	300 mA	400 mA	500 mA
Time	4/5	2/5	'/5	2/15	'/10	'/12
1.						
2.						
3.						
4.						
5.						
6.						
7.						
8.						
9.						
10.						

CONTROL CHART FOR GENERATOR TRACKING STEP WEDGE

inch FFD
Table top

kVp
mAs

mA Stations Time	50 mA	100 mA	200 mA	300 mA	400 mA	500 mA
1.						
2.						
3.						
4.						
5.						
6.						
7.						
8.						
9.						
10.						

CONTROL CHART FOR TRACKING GENERATOR PHANTOM

Type of phantom

inch FFD
grid

kVp
mAs

mA Stations Time	50 mA	100 mA	200 mA	300 mA	400 mA	500 mA
1.						
2.						
3.						
4.						
5.						
6.						
7.						
8.						
9.						
10.						

SYSTEM NORMS

Overview

Every type of x-ray system is different and includes some variables (discussed in earlier sections). The patient, expertise of the technologist, diagnostic examination, radiographic processing, and preference of the radiologist are a few of the variables that influence the diagnostic outcome of any radiographic procedure.

The following is an example of a system norms chart for cineradiography.

Radiographic Functions:	*Patients*		
	Children	*Average Adult*	*Large Adult*
kV	50 to 60	70 to 80	Low as possible
mA		As high as possible for all groups	
Pulse width	1 to 3	3 to 5	Greater than 5
Frame rates	30 to 60	30 to 60	30
F-stop		3.0 to 4.0 (avg. 3.5)	
Microroentgens		20 to 40 (avg. 25)	
Developer temperature		Manufacturer's instructions (avg.) 80°F	

X-RAY CINE FILM SYSTEM

Overview

This section deals with the x-ray system, image intensifier system (I.I.), optics cube, and the cine camera to produce the optimum cine film.

GENERAL STATEMENTS

1. X-ray to the cine camera system is equal to the x-ray system.
2. Cine camera system equals photographic system.
3. In direct x-ray systems, only the radiation dosage controls the film density.
4. In cine-radiography film, density is not only dependent upon the radiation dosage but also upon the F-stop of the camera.

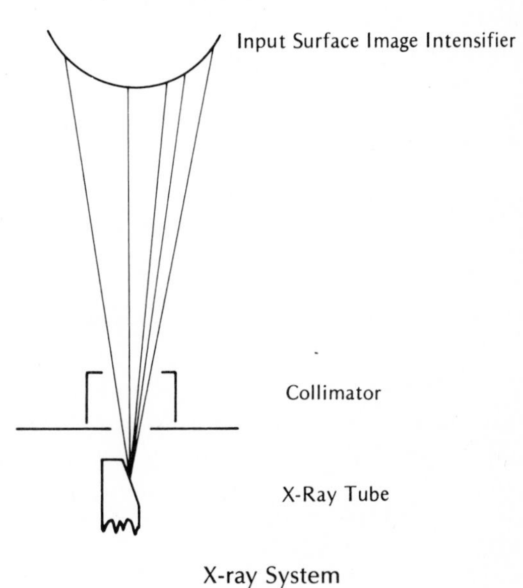

X-ray System

I. X-RAY SYSTEM

The initial step in the imaging system is the process of placing the x-ray image on the face of the image intensifier. The quality of this image is dependent upon the component parts of the x-ray cine film system.

The focal spot of the x-ray tube is the starting point of the x-ray system. It is the first component that will cause a loss of resolution. A focal spot is the effective area that x-ray radiation is generated from within the x-ray tube. Different-sized focal spots measured in millimeters of mercury result in varying degrees of resolution. The larger the focal spot, the greater the power and the poorer the resolution. The smaller the focal spot, the greater the resolution and the poorer the power. (*Examples:* small focal spot used to view a hand and a large focal spot used to view barium in the stomach.)

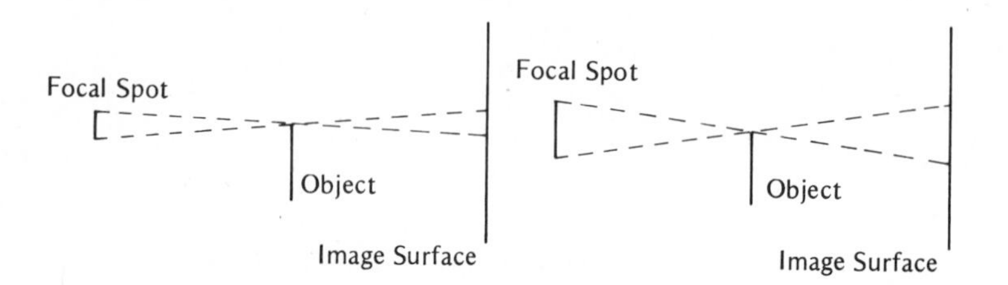

The smaller the focal spot, the lower the kW rating that the x-ray tube can handle. The kW rating is a product of kV, mA, pulse width, and the time or length of cine run.

The kW ratings of all x-ray tubes, to this date, do not allow conventional cine filming on 0.1-mm focal spots. 0.6-mm focal spots are limited in kW ratings. Only very small adults and children can be filmed on a 0.6-mm focal spot. The majority of cine cardiac filming is presently done on a 1.2-mm focal spot. An interesting point to remember is that a new x-ray tube may have 0.6-mm by 1.2-mm focal spot printed on the x-ray tube and associated paper work with it. However, the actual focal spot size may be larger or very much larger than stated. Consult the specifications as listed on the paperwork that comes with the tube. The specifications will state what tolerance is allowed for an actual focal spot and still let it be listed at a specific size.

EXAMPLE

Listed size: 1.2 mm
Tolerance: −20% to +50%
Size limit within tolerance: 0.96 to 1.8 mm

Unfortunately for the user, many tubes are very close to their maximum tolerance. Also, a tube's focal spot can increase, and usually does, with age (use). If the tube is run at, or close to, its maximum mA rating, the filament can warp. This causes an increase in the focal spot size from two to three times the listed size. There are various methods for testing focal spot size. Two such methods are the lead star and the pin hole camera.

A. Geometry of the x-ray Sytem

Geometry also will cause a change in the x-ray resolution. The closer the object is to the imaging input surface, the better the resolution. Viewing a film with a straight edge of good radiographic contrast and moving this edge away from the input surface results in the blurring of the straight edge as well as magnification of the edge. Due to this magnification, a greater number of line pairs per millimeter are visualized than if the object were close to or on the input imaging surface. A line pair consists of two lines, one of which is radio-opaque and one which is radiolucent.

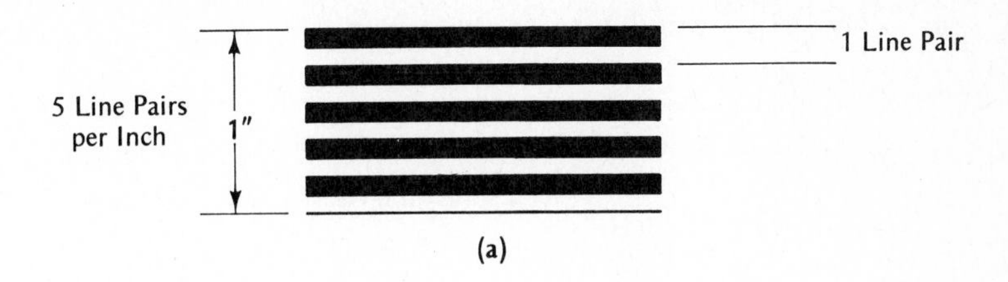

(a)

While viewing a line pairs phantom on a television monitor, move the phantom from the surface of the image intensifer (I.I.) away until a distance is reached halfway between the I.I. and the x-ray tube. (Always use lead apron and gloves during this procedure.) Note that when the phantom is next to the I.I., the image is sharp and no less than 50 line pairs per inch or 2 line pairs per millimeter are visible. As the phantom is moved away from the I.I., the image loses its sharpness, but more line pairs seem to be visible. Magnification also increases.

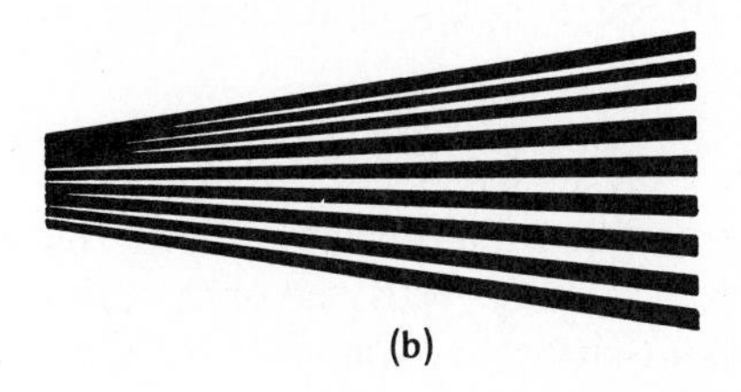

(b)

All cardiac catheterization laboratories and special procedure laboratories should have their own line pairs phantom for weekly quality control procedures. Thereby, subtle changes are picked up and corrected before major problems develop.

Technique

The radiographic technique of low kV, high mA may be one of the most misunderstood procedures in cine radiography. The kV wave form must be square or the kV does not change more than 1% to 2% over the duration of a single pulse (constant potential generator). In a constant potential generator, the limiting factor is the x-ray tube. In generators other than constant potential, the limiting factor is the specification of the generator itself. In these generators, the mA should not be run higher than that specified by the manufacturer. The reason for this is that, during a single pulse, the kV is dropping. The higher the mA, the faster and lower the kV drops. The disadvantage of these generators is that as the kV drops the amount of light to the film drops at a very fast rate.

EXAMPLE

If a pulse starts at 80 kV and drops to 78 kV over a pulse width of 3 ms, an analysis of kV to the amount of light reaching the film would show that approximately 70% of the total light to the film is in the first 0.5 ms. In the last 0.5 ms, the yield is less than 10% of the total light to the film. The solution to this problem is limited experimentation. Generally, the best procedure to follow is the manufacturer's recommendations. In summary, keep mA high but within the specifications of the equipment. This will yield good contrast and kV sufficient to expose the film to the desired level.

II. IMAGE INTENSIFIER (I.I.)

This may be referred to as an image amplifier. The component parts of the image intensifier are as follows:

1. Input Surface:
   a. Glass envelope
   b. Converter coating (converts an x-ray photon to light)
   c. Photocathode (the light from the converter coating causes the photocathode to release electrons)
2. Internal Elements:
   a. Grid I: focusing element
   b. Grid II: secondary focusing element
   c. X-ray shield
3. Output phosphor
4. High voltage supply: supplies proper voltages to internal elements and output phosphor.

The x-ray image penetrates the glass envelope, which causes the converter coating to generate the image in light. The photocathode is in direct contact with the converter coating. Thus, the photocathode captures the light image and releases this image in electrons. The electrons are then accelerated and focused by grid I and grid II to an electron image on the output phosphor. This image is a highly intensified image. The intensification is caused by the acceleration of the electrons by the high voltage. When these accelerated electrons strike the output phosphor, they generate thousands of times more light in the output than there was light from the input converter to the photocathode. The x-ray shield is to prevent any x-ray photons from reaching the output phosphor. This results in an image of sufficient light level for exposure of light sensitive film (35mm cine film) or for an image on a TV camera.

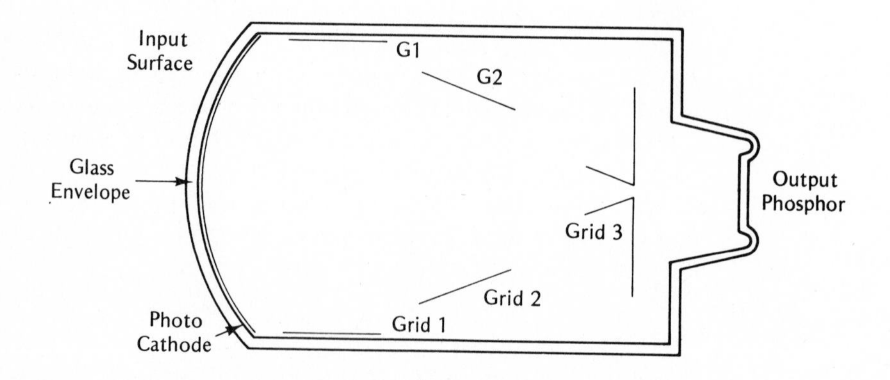

Problems of an Image Intensifier

These problems include but are not limited to (1) lack of proper focus, (2) wrong image intensifier for the application, and (3) image intensifier that is worn out. Focusing the II may be done by the laboratory's radiographic service engineer. There are two focusing adjustments. One is for the center portion of the II and the other is for the perimeter. Both of these foci interact and overall focus is a compromise of the two at best.

In the cardiac catheterization laboratory, the greatest information is obtained in the center 70% of the image. The center portion of the image thus should be enhanced in resolution, within reason, at the expense of the perimeter of the image.

When purchasing a new image intensifier, for a nominal charge the II can be selected for either high resolution or high contrast, but not both. In cardiac catheterization work, the slight deterioration in resolution (100 line pairs/inch) encountered when selecting for high contrast does not create a problem. A 100 line pairs/inch will allow visualization of third and some fourth orders (branches) of the coronary arteries. That in itself is of little use but is a good indication of how accurate a coronary occlusion can be measured in the viewing area.

III. OPTICS CUBE

The optics cube is comprised of four components: (1) II output (columinating) lens, (2) beam splitter, (3) television lens, and (4) cine lens. These light handling components are either in the optics cube or sticking into it.

Optics Cube

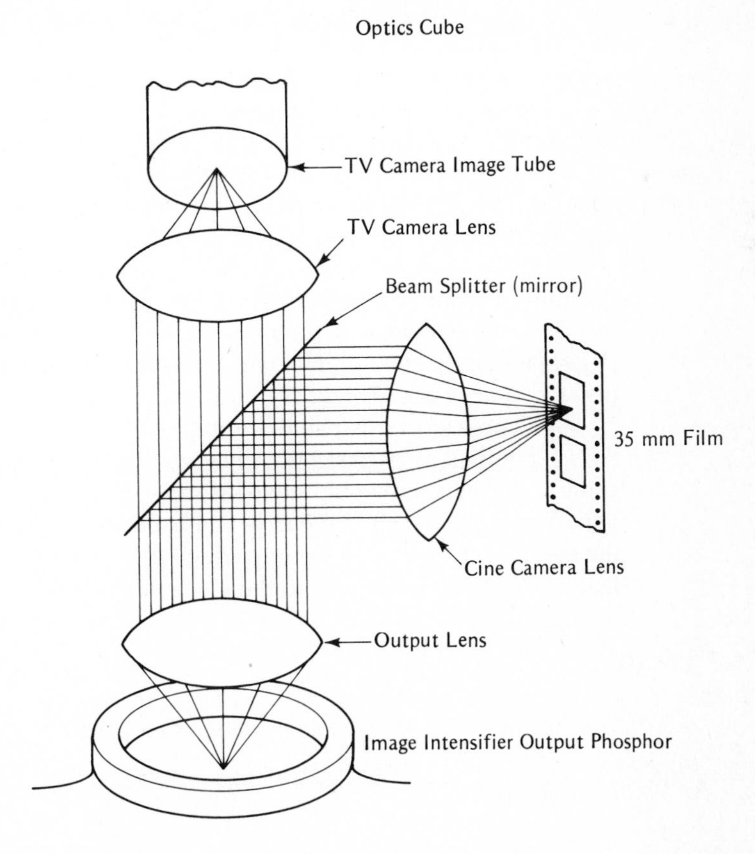

The image intensifier output lens is focused at the factory and **should not** be touched in the field. The radiographic service engineer does not have the equipment or the expertise to focus this lens correctly in the laboratory. If you have ever looked through a telescope at the stars, you know that when you view the planets, the telescope needs to be refocused. Thus, it is not hard to imagine that an alignment system within the confines of the cardiac catheterization laboratory must be found. If the I.I. output lens is slightly out of infinity focus, the TV lens and the cine camera lens can correct for it and still give the same resolution.

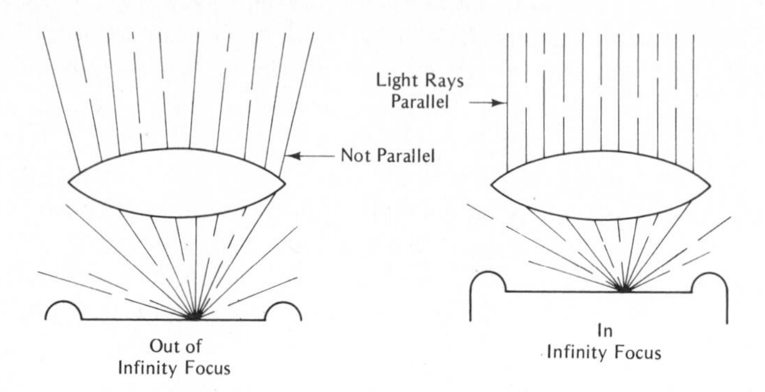

The loss of light due to the lens being out of infinity focus (light rays from the I.I. output lens are not absolutely parallel) occurs because the maximum amount of light available does not enter the next lens (television or cine), and thus some light is lost before it gets to the film. When the I.I. lens is lightly out of infinity focus, the loss of light is insignificant. However, if the lens is badly out of focus, the amount of radiation to expose the film must be increased. If the generator is at full mA, then the kV must be increased and, consequently, the contrast starts to deteriorate.

The beam splitter or bounce mirror is used during cine-radiography. It is a front surface mirror that reflects approximately 90% of the light to the cine camera, and the remaining 10% passes through the mirror to the TV camera. Thus the name beam splitter. During fluoroscopy, the mirror is motor driven out of the way. The unrestricted light from the I.I. output lens goes directly to the TV camera.

As long as the image is centered on the film, the beam splitter mirror is functioning properly. Like any mirror, it must be cleaned. Dust, dirt, or a fingerprint will deteriorate focus and contrast. This will result in a loss of light to the film. As stated previously, the bounce mirror (beam splitter) and lenses are all in or sticking into the optics cube. The optics cube is light tight but not dustproof.

The optics cube is painted a flat black to absorb extemporaneous light from the I.I. output lens. If the optics cube was not painted a flat black and had a shiny surface, the light rays would strike the shiny surface and bounce into the cine camera lens. This would result in the cine film becoming fogged by the light. The next time the equipment is serviced by the radiographic service engineer, ask to see the optics cube, which is located next to the cine camera.

IV. CINE CAMERA

The 35mm cine camera is a very important part of the overall system.

A. Camera Shutter

1. The time the shutter is open varies with the framing rate of the cine camera. [*Example:* 60 frames/second (FPS), shutter opening time 6 ms; 30 FPS, shutter opening time 12 ms.]

2. The beginning and end of the exposure light to a frame of film is controlled by the **x-ray generator** and not the shutter. If the shutter is in the field during the exposure, it can cause a shadowing effect on each frame.
3. The shutter is helpful in alternate exposures of biplane systems by preventing light generated from scatter radiation exposing the film of the other plane during its film advance cycle.
4. Some shutters on cameras rotate and others have a slide action.

B. Lenses

1. Lenses are compound (more than one elements).
2. Each element of the compound lenses is precision ground and made of very high optical grade glass. This results in the compound lens system being relatively expensive.

C. Camera Lens Focus

1. Image intensifer output phosphor to film results in a two-dimensional image.
2. Using a mirror to redirect light rays. **Does not** change the focus of the lens.

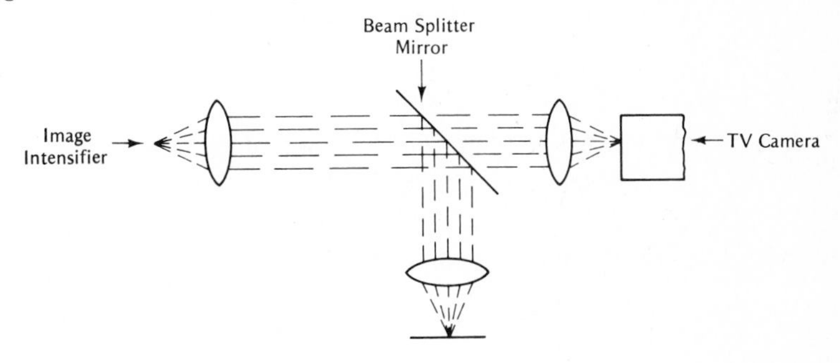

D. Camera framing

Camera framing deals only with the size of the image.

Framing Mode	*Film Area*	*Focal Length of Lens*
Underframing (18 mm)	58%	75 mm
Middle framing (24 mm)	88%	85 mm
Overframing (30 mm)	100%	100 mm

E. Lens Vignetting

This condition is a fall off of light around the edges of the film.

F. Lens Aperture

The lens aperture controls how much light is going to get through to the film.

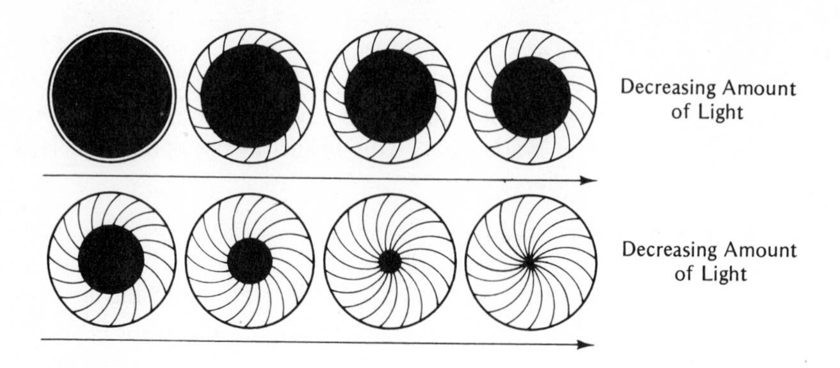

G. What Is an F-stop?

1. An F-stop is the focal length of the lens (dictated by the manufacturer when focused at infinity focus) divided by the diameter of the opening (aperture).
2. Focal length = F
   Diameter = D
   F-stop = F/D
3. Example
   a. 100-mm lens with lowest F number of 2 equals focal length of lens is 200.
   b. Close aperture to 71 mm equals F-stop of 2.8. (The light to the film has been cut directly in half.)
4. The F-stop should be wide open or closed down only 1 F-stop. (Any further closing down of the F-stop will cause an increase of radiation.)

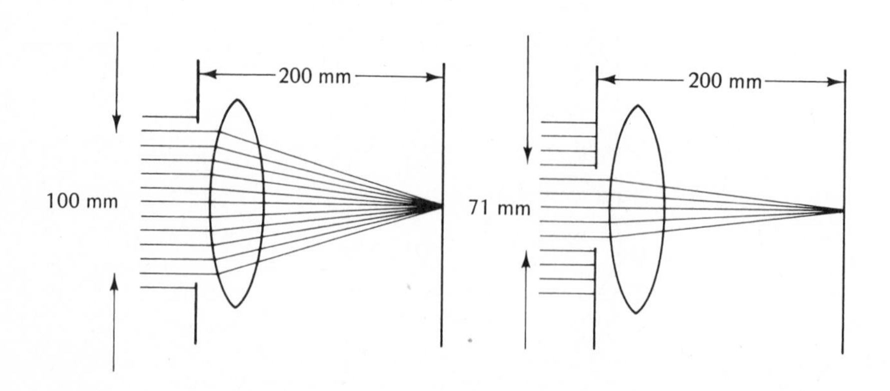

Some systems have been set up with the F-stop at mid-position. *Example:* a 100-mm F2 cine lens set at an F-stop of 5.6. The radiation must then be adjusted to give proper exposure of the film.

This latter method is wrong for the following reasons:

1. Unnecessary excessive radiation dosage to the patient.
2. Loss of contrast due to higher kV.
3. Scatter radiation to laboratory personnel up two to three times higher than normal.
4. Due to higher average kV, there will not be enough reserve kV for the "heavy" patient.

H. Camera Lens F-stop

The camera lens F-stop varies and measures the amount of light hitting the film.

1. F-2.8 and F-2 = 2× difference (factoral stop system)
2. F-stop = F/D
3. One F-stop = change by factor of 2
   $2 \times \sqrt{2}$ = 2.8
   2 × change = 0.3 log of exposure
4. D = F/F-stop

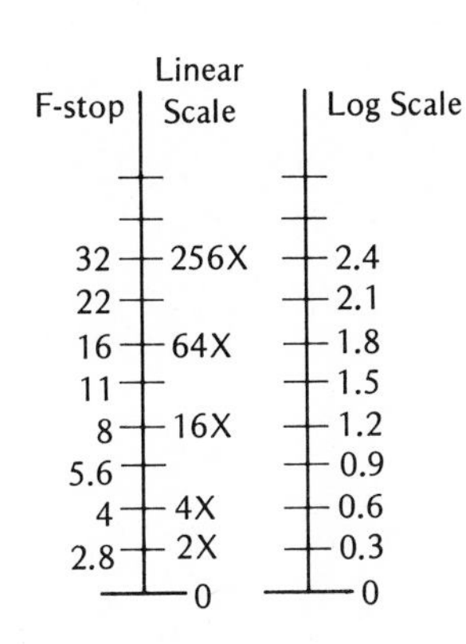

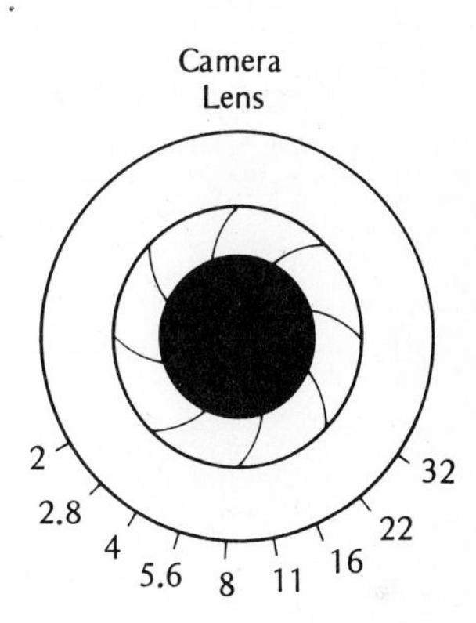

I. Calculation of Lens Aperture Diameters and Areas

1. Area = $\pi(D/2)^2$
   Example: If D = 20, 3.1414(10) = 314.16 mm^2
2. Diameter = $\sqrt{4A/\pi}$ (A = area)
   Example: If A = 267, D = 4(267)/3.1416 = 18.4

J. Summary of Formulas

1. *F-stop* = F/D
2. D = F/F-stop
3. Area = $\pi (D/2)^2$
   Diameter = $\sqrt{4A/\pi}$

K. Fine Tuning and Optimum Film Quality

1. Due to the fact that in cine-radiography the film density is not only related to the radiation dosage but also to the F-stop, a change of 1 F-stop may result in better films.

2. If you are going to change the radiation dosage, you must change the F-stop to keep everything in synchronization.

L. Maintenance of the Cine Camera

1. Keep the film gates of the camera clean.
2. Take time in loading the magazine to prevent jams. The magazines are very expensive ($5,000 to $8,000).
3. Have the service engineer, during routine maintenance, clean and adjust the magazines.
4. Clutches should be cleaned and care taken not to harm them.

V. CINE-RADIOGRAPHIC FILM

High contrast, high speed, good latitude low grain content, and excellent consistency reel to reel and lot to lot are just a few of the characteristics that one looks for in selecting cineradiography film. A few other particular points to look for are that it does not leave excess emulsion in the camera, process too fast, nor collect dust or scratch easily. All these items are to be negotiated between the director of the laboratory, his or her catheterizing cardiologist, and the film company. It must be understood from the beginning that all these factors interact and that the film must be compatible with the x-ray cine system.

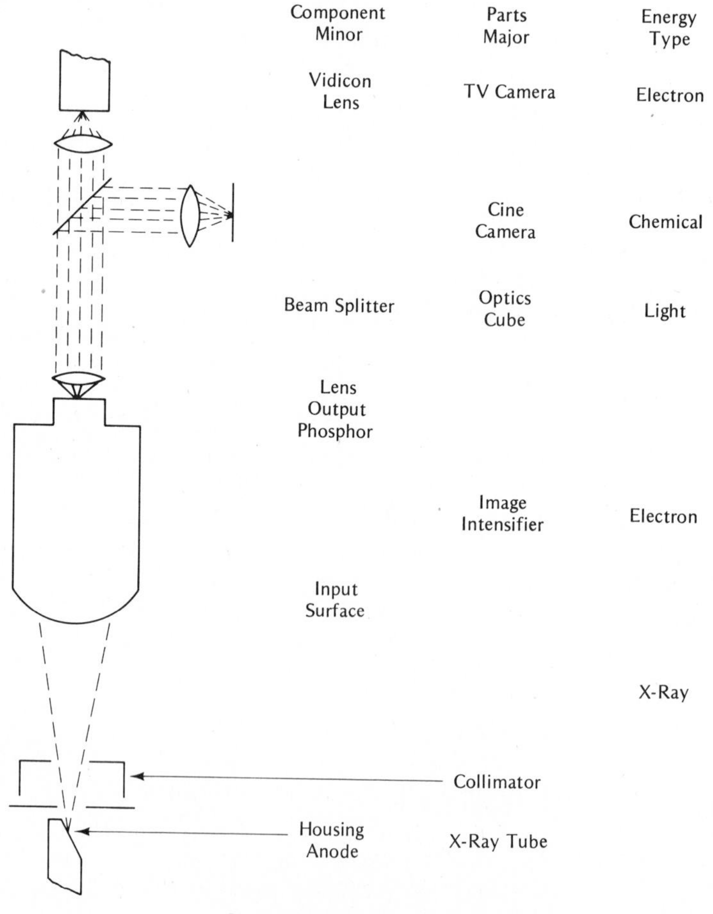

X-ray Cine Film System

CONTROL CHARTS

Overview

Documentation is an important aspect of quality control and the quality assurance program. This documentation is completed in several steps. One of the modes of recording the results is the use of control charts.

WHAT IS A CONTROL CHART?

A control chart is a graphical means of recording the variations of the level of a specific process with time. The established operating standard or level is designated by OS, the upper control limit is designated by UCL, and the lower control limit is designated by LCL. The vertical axis signifies the item being measured and the horizontal axis signifies time.

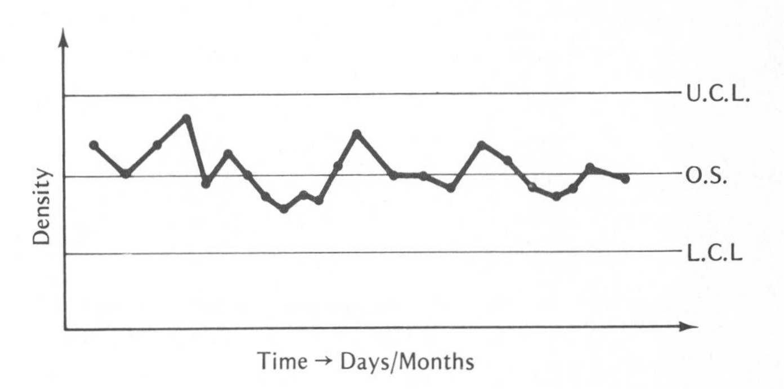

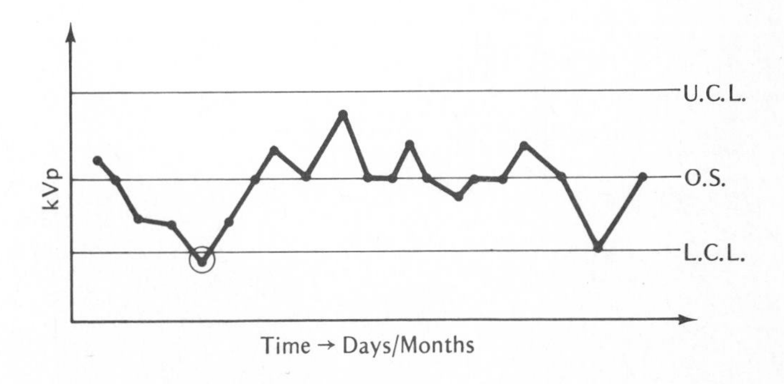

OBJECTIVE

To provide a daily graphical means of documenting (recording) data for easy analysis and interpretation of established operating standards.

PLAN: GATHER EQUIPMENT

1. Specialized charts, graph paper
2. Ruler
3. Pen, pencil, miscellaneous

Implementation	*Rationale and Key Points*
1. When setting up control charts, established operating levels, and upper and lower limits must be determined. (See chart on page 131.)	1.1. This allows for easy analysis of what is actually happening on a daily or monthly basis. 1.2. These parameters will determine when corrective action must be taken.
2. Collect all data to be plotted on charts. a. Date b. Data points c. Comments d. Corrective action if appropriate	
3. Determine if any of the data points lie outside of control limits.	3.1. If this is the case, before initiating corrective action, recheck calculations and, if needed, the process.
4. If corrective maintenance has been performed, repeat quality control.	4.1. This is an important step, especially if a contract service is used.
5. Plotting of the control points: a. Any data point out of the control limits should be circled. b. Connect data points with a pencil.	5.1. By circling the out of control data points, they stand out. 5.2. By connecting the data points, a trend, drift, or normal readings may be visualized.
6. Observe for any marked trend or drift on the control chart. A trend or drift is indicated when three or more data points move in the same direction.	6.1. A trend or drift in one direction or another usually indicates a need for corrective action.
7. Record any corrective action taken under Remarks.	
8. Random variation on the control charts is normal.	8.1. Random variation is to be expected. It is caused by technological error and normal minute variations in the process being monitored.

PROBLEMS AND SOLUTIONS

Problem: Quality control charts not kept up to date.

Solution: Quality control must be approached as a team effort. Job descriptions and responsibilities must be clearly delineated.

Problem: Failing to recognize random variation.

Solution: In-service education. Explanation that random variation is normal, that the data points should fall around the established operating standard level and be confined within the control limits.

Problem: Placing quality control data on many small pieces of paper kept in various places.

Solution: Develop an organized record system and keep all the quality control equipment, records, and manuals in one area.

Problem: To institute corrective action when a trend or drift begins to occur or data points are out of control.

Solution: Make everyone aware of the procedure to be followed when corrective action must be instituted.

Responsibilities	*Action*
1. Quality control coordinator or designee	1.1. Is responsible for daily quality control and recording the results.
	1.2. Initiates corrective action when necessary.
2. Chief technologist	2.1. Overviews the quality control program and follows through with appropriate action.

APPROVED: ______________________

Date: ______________________

ITEMS TO BE INCLUDED ON RADIOGRAPHIC PROCESSING CONTROL CHARTS

1. Radiographic processor number
2. Month and year
3. Date scale
4. Graphs
   a. Base plus fog
   b. Medium density
   c. High–low density
5. Replenishment rate
   a. Date
   b. Developer rate
   c. Fixer rate

6. Temperature control
   a. Date
   b. Developer
   c. Water
   d. Dryer
7. Comments
   a. Preventive maintenance
   b. Corrective maintenance
   c. Effectiveness

X-RAY PROCESSING CONTROL CHART

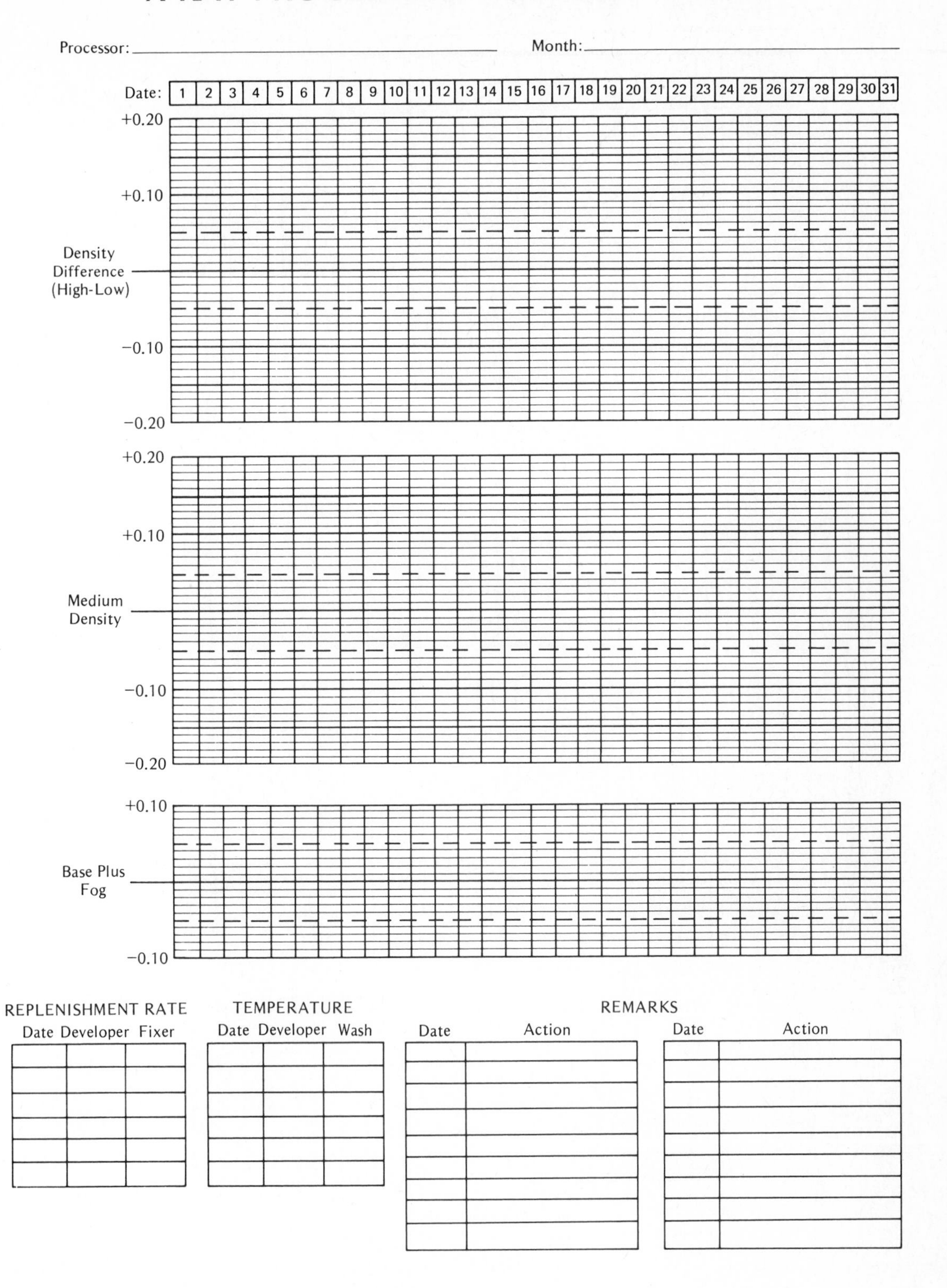
Processor: ______________________ Month: ______________________

Date: 1 2 3 4 5 6 7 8 9 10 11 12 13 14 15 16 17 18 19 20 21 22 23 24 25 26 27 28 29 30 31

+0.20
+0.10
Density Difference (High-Low)
−0.10
−0.20

+0.20
+0.10
Medium Density
−0.10
−0.20

+0.10
Base Plus Fog
−0.10

REPLENISHMENT RATE

Date	Developer	Fixer

TEMPERATURE

Date	Developer	Wash

REMARKS

Date	Action	Date	Action

Courtesy of E.I. duPont de Nemours & Co., Inc., Wilmington, Del.

IMAGING MANAGEMENT QUALITY CONTROL PROCESSING CHART

FORM E-22121

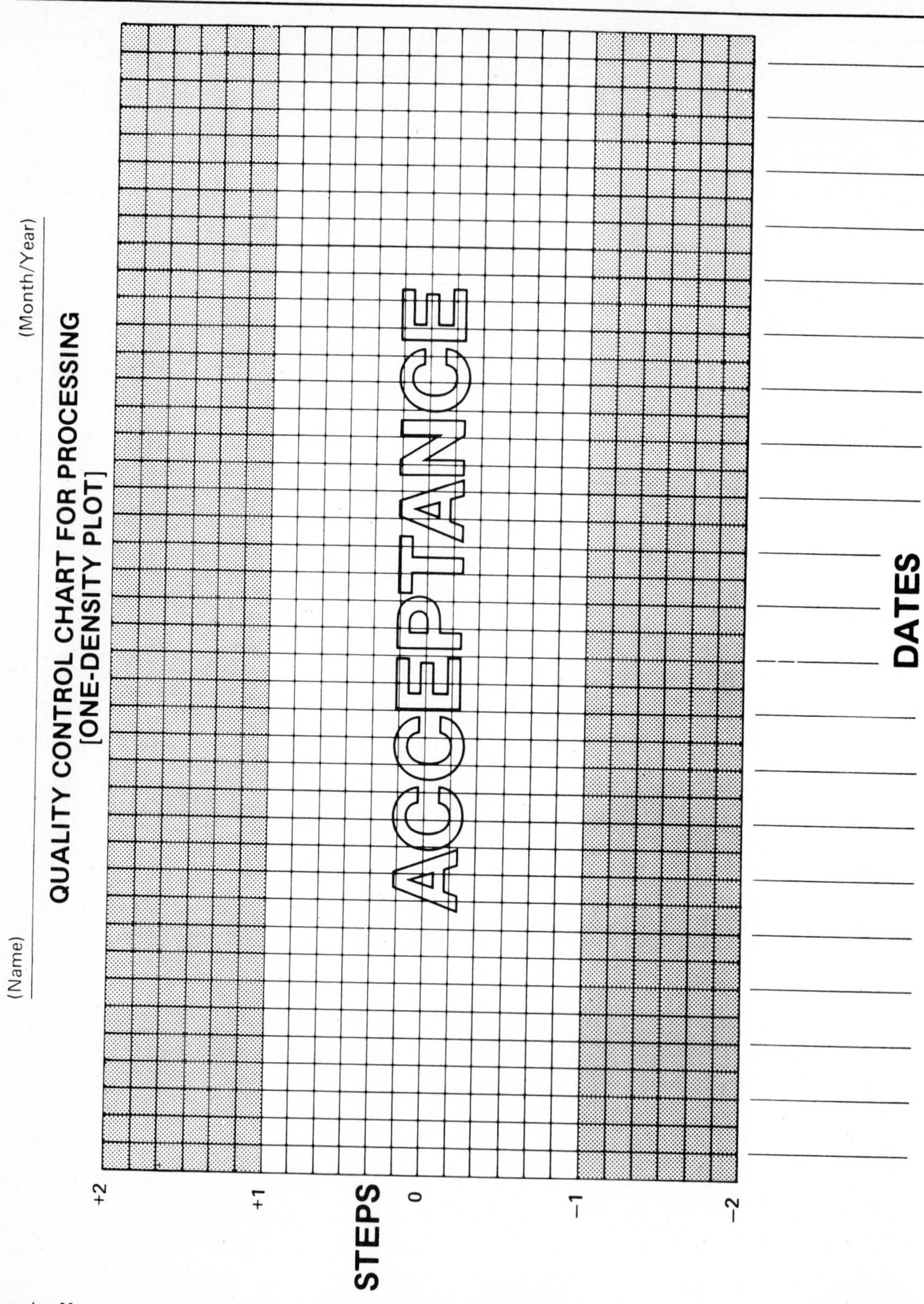

Imaging Management puts you in control of quality, costs and patient care.

Printed in U.S.A. R-8-78

IMAGING MANAGEMENT

CHEMICAL ANALYSIS REPORT

FORM E-22285

Date Sampled:
XRTS Received:

TO: R.J. Trinkle, XRTS CUSTOMER:

FROM:

DESCRIPTION OF PROBLEM

TEST CONCLUSIONS

SYSTEM DATA

FILM	A	B	**Our Normal Recommendation**
Type			
Emulsion No.			
CHEMISTRY			
Developer Type			
Fixer Type			
PROCESSOR	A	B	
Location			–
Model			–
OPTIMUM DEVELOPMENT – 15.0 Bit Total	BITS	BITS	
Immersion Time (sec)	=	=	–
Temperature (°F)	=	=	–
Developer Type	=	=	XMD
Bit Totals			15.0 Bits
REPLENISHMENT			
Developer (cc/14" Average)			60
Fixer (cc/14")			90
Wash (gal/min)			1 to 2

Imaging Management puts you in control of quality, costs and patient care.

Printed in U.S.A. R-8-78

TEST RESULTS

CHECK LIST NEEDED	√	A	√	B	COMMENTS	Standards
SENSITOMETRY (Send Processed Sensitometer Exposure)						
Bit Speed						
Toe Gradient						
Av. Gradient						
Step 21 Density						
Base + Fog						
FILM ARCHIVAL QUALITY (Send Processed Sensitometer Exposures)						
Hypo Retention (μ gm/in²)						0 to 25
Silver Retention (μ gm/in²)						0 to 25
CORRECT MIXING (8 ounces are essential)						**XMD/XMF**
Dev. Repl. Sp. Gr.						1.077 ± .004
Fixer Repl. Sp. Gr.						1.110 ± .004
Developer Du Pont?						Yes or No
Fixer Du Pont?						Yes or No
CORRECT REPLENISHMENT (Send Used Developer or Fixer)						
Developer K Br (gm/L)						8.0 ± 2.0
Fixer Silver (Tr. Oz/gal)						.8 ± .2
CHEMICAL ACTIVITY (Full Bottles Prevent Oxidation)						**XMD/XMF**
Developer						
pH						10.3 ± .1
Hydroquinone (gm/L)						20 to 25
Oxidized HQ (gm/L)						18 to 10
Fixer in XMD						0
Developer Replenisher						
pH						10.5 ± .1
Hydroquinone (gm/L)						>25
Oxidized HQ (gm/L)						<10
Fixer						
pH						4.4 ± .1
Thiosulfate (gm/L)						
Fixer Replenisher						
pH						4.2 ± .1
Thiosulfate (gm/L)						
OTHER TESTS						

Courtesy of E.I. duPont de Nemours & Co., Inc., Wilmington, Del.

ESTABLISHMENT OF RADIOGRAPHIC PROCESSOR CONTROL LEVELS QUALITY CONTROL PROGRAM

Overview

Before any type of quality assurance program can be established, one must develop quality control norms for each individual piece of equipment and/or procedure to be used in that program.

Objective

To determine operating levels for each radiographic processor in the radiology department. Operating levels should be optimized so that each processor operates within the same limits.

PLAN: GATHER EQUIPMENT

1. Sensitometer
2. Densitometer
3. Control films
4. Control charts
5. Thermometer
6. Stop watch
7. Control emulsion
8. Fresh chemicals
9. Miscellaneous writing tools

Implementation	*Rationale and key points*
1. Drain processor tanks and flush with water.	1.1. Process ensures that there will be no contamination.
2. Replace recirculation developer filter.	2.1. Ensures that there will be no contamination of new chemicals.
3. Drain and flush replenishment tanks.	3.1. Ensures that the replenishment tanks will be free from contamination.

Implementation	*Rationale and key points*
4. Mix fresh developer and fixer for the replenishment tanks according to the manufacturers' directions.	4.1. Ensures that the chemicals will be mixed correctly.
5. Refill replenishment tanks with newly mixed chemicals.	
6. Start the replenishment pumps and clear the lines of any water.	6.1. Water in the lines between the replenishment tanks and the processor would contaminate the new chemicals.
7. Flush the processor tanks with water for the second time.	7.1. Ensures that the processor tanks are free from contamination.
8. Fill the developer and the fixer tanks with fresh chemicals.	8.1. Mix fresh chemicals following the manufacturers' directions.
9. Let the tanks sit for 20 to 30 minutes. Drain the tanks.	9.1. This process is known as seasoning the tanks.
10. Refill the developer and the fixer tanks from the replenisher tanks.	10.1. The tanks should be free from any contamination and ready for use.
11. Replace the fixer and developer racks.	
12. Replace the crossover racks.	
13. Allow the processor to equilibrate for 1 hours.	13.1. This equilibration period allows for adequate mixing of the chemicals. It also allows for the correct temperature to be reached.
14. Check *all* the temperatures of the solutions. a. Developer b. Fixer c. Water	14.1. Set the temperatures at the levels specified by the processor and film manufacturers.
15. Check the replenishment rates.	15.1. Replenishment rates specified by the manufacturer are always set high. One may lower the developer replenishment rate. Monitor the results by the sensitometer strips. 15.2. Leave the fixer replenishment *alone*. This rate may have to be increased if there is a significant increase in patient load.
16. Check the film transport time with the stop watch.	16.1. Transport time is the time interval from when the leading edge of the film enters the processor until the leading edge of the film exits the dryer.

Implementation	*Rationale and key points*
17. Utilize the processor until approximately 1 gallon of replenishment chemicals has been used. This is about fifty 14 × 17 films.	17.1. This period of time is an equilibration period for the processor. During this time period, check the replenishment pump function and the overall processor function.
18. Expose six sensitometric strips.	18.1. Expose these strips according to the manufacturers' directions.
19. If the film is double emulsion film, expose both sides of the film.	19.1. In this case only three films need to be used.
20. Process these strips after the processor has been utilized for 1 hour.	20.1. See item 17.1.
21. Process the sensitometric strips by the thin edge leading and using the same side of the tray.	21.1. Every process must be done in the same manner to ensure reproducible results.
22. Zero and calibrate the densitometer.	22.1. The densitometer must be calibrated in order to ensure correct readings.
23. Read the densities of the six strips. Take care that each strip is read in the center of the strip.	23.1. False readings may be obtained if the strip is read near its edge.
24. Determine the average densities of the six strips for each step. a. Base plus fog b. Mid-density c. Density difference	24.1. Add the six strip steps and divide by 6 to obtain the average base fog, mid-density, and the density difference.
25. Document the steps producing the densities nearest to 0.25, 1.0, and 2 above the base fog.	25.1. These numbers will determine the control limits.
26. Determine the appropriate data points. Plot these points with the control limits on the control charts.	26.1. The control charts for each radiographic process have now been established.
27. Adjust all the radiographic processors using the *same* type of film and chemicals.	27.1. This is necessary for reproducibility of results between all the processors.

Responsibilities	*Action*
1. Quality control coordinator or designee	1.1. Establishes the control for each new film batch.

APPROVED: ______________________

DATE: ______________________

Courtesy of E. I. duPont de Nemours & Co., Inc., Wilmington, Del.

DAILY RADIOGRAPHIC PROCESSOR QUALITY CONTROL

Overview

The quality control techniques for checking radiographic processing are an important daily task for the quality control coordinator in the radiology department.

OBJECTIVE

To ensure that all radiographic processors are operating on the same level, producing high quality film, and are within the established control standards. This is done on a daily basis.

PLAN: GATHER EQUIPMENT

1. Sensitometer
2. Densitometer
3. Thermometer
4. Control Emulsion
5. Control Charts
6. Miscellaneous writing tools

Implementation	*Rationale and key points*
1. Turn processors on early in the morning to let them equilibrate.	1.1. Allow sufficient time for the processor to stabilize temperature and for mixing of the chemicals.
2. Follow the manufacturers' start-up procedures for the processors.	2.1. Each manufacturer has a slightly different start-up procedure.
3. Check these items in order. a. Solution temperatures b. Dryer temperature c. Water flow rate d. Replenishment rate	

Implementation	*Rationale and key points*
4. Run four to six clean-up sheets through the processor.	4.1. The clean-up sheets are run to remove any residue from the racks and to check for any scratches.
5. Expose enough sensitometric strips for a single day. a. Single-side emulsion film, expose on emulsion side. b. Dual-sided emulsion film, expose on both sides.	5.1. If the sensitometer only exposes one side of the film, it will be necessary to expose both sides
6. Let the strips sit for 30 minutes before processing. Do not let more than 4 hours elapse before processing the strips.	6.1. This process decreases the variability from strip to strip.
7. Processing of strips: a. Two per machine b. Thin end leading c. Same location on processor feed shelf.	7.1. These techniques ensure reproducibility.
8. Zero and calibrate the densitometer.	8.1. The densitometer must be zeroed and calibrated to ensure accurate readings.
9. Read the three density levels including the base fog level. a. Average the two exposures on the dual emulsion film. b. Plot the base plus fog, mid-density, and density difference on the control charts.	
10. Observe the control chart.	
a. Are all the data points within the control limits?	10.1. Circle all the control points out of control.
b. Is a drift or trend apparent?	10.2. Three to five data points in the same direction establishes a trend or drift.
c. Is random variability present?	10.3. Random variability is a normal occurrence in control charts.

CORRECTIVE ACTION

1. Reasons for corrective action:
   a. Data points outside control limits.
   b. Three to five data points headed in the same direction (trend, drift).
   c. Three to five data points are consistently above or below the aim points.

2. Implementation for corective action:
   a. Recalibrate the densitometer.
   b. Reread the control strip to eliminate a reading error.
   c. Examine the processor parameters.
      (1) Temperature of developer, water, dryer
      (2) Replenishment rate
      (3) Transport speed
      (4) Agitation
   d. Redo the control monitoring strip.
   e. If the control monitoring strip remains out of control limits, institute the trouble-shooting procedure for radiographic processors.

Responsibility	*Action*
1. Quality control coordinator or designee	1.1. Completes daily quality control and documents the results.
2. *Chief Technologist*	2.1. Assists in the process of trouble-shooting quality control problems when the need arises.

APPROVED: ______________________

DATE: ______________________

PROCESSOR REPLENISHMENT

Overview

This is a difficult parameter to set on any processor system. General guidelines are offered by the manufacturers of the film and by equipment and chemistry manufacturers. However, these guidelines need to be reviewed and approved by your quality control coordinator and committee.

As one becomes involved in quality control, one will find that many variables will influence the replenishment rate. The two most common variables are the number of films processed over a particular time period and the types of films processed.

OPTIMIZATION GUIDELINES FOR REPLENISHMENT RATES

1. Set initial replenishment rates as specified by the manufacturer.
2. Many radiographic processors can run at a lower developer rate of replenishment than specified by the manufacturer. Do not hesitate to reduce the developer rate over a period of time, as long as one remains within the operating parameters on the sensitometric control monitoring strips.
3. **Always** maintain your fixed replenishment at the level specified by the manufacturer. This level may have to be increased but should not be decreased.
4. Most flow meters associated with the developer replenisher and water are not precisely calibrated. Thus replenishment rates between processors may be different, even though they may be the same type of processor. The **only** method to ensure proper replenishment rates is by the quality control processor charts.
5. **Remember**, in processors that handle a low volume of film an increase in the replenishment rate is needed to produce high quality radiographs. This is known as "flood replenishment."
6. It is recommended to shut down a processor if it is not used for periods of 2 to 3 hours at a time. This will prolong the life of the radiographic processor and decrease the amount of evaporation and oxidation of the developing solution.
7. An increase in the volume of work that a processor does in a given period of time changes the replenishment rate.

FLOOD REPLENISHMENT FOR THE LOW VOLUME RADIOGRAPHIC PROCESSOR

Overview

Flood replenishment is a modified means of radiographic processor replenishment for low-volume work. This may include cine, nuclear, mammography, and ultrasound. The principle is to over replenish to compensate for the lack of volume in films and still obtain optimum quality physical image radiographs.

OBJECTIVE

To maintain high-quality films for low-volume processors.

PLAN: GATHER EQUIPMENT

1. Five-minute, 120-volt interval timer
2. Strain-relief wire cable connectors
3. Six-foot-length, four-conductor electrical cable
4. Qualified service engineer or a qualified electrician

Implementation	*Rationale and key points*
1. Have the timer installed by an x-ray service engineer or qualified electrician. Replenishment pump should operate for 20 seconds out of every 5 minutes.	1.1. Ensures that there will be approximately 780 ml of developer and fixer solution replenishment for each hour that the processor is operational.
2. Flush radiographic processor system of all chemicals after timer has been installed.	2.1. Removing and cleaning processor and replenishment tanks of chemicals will ensure that the chemistry is fresh and uncontaminated for initiating flood replenishment.
3. Remove the old developer recirculation filter and replace it with a new one.	3.1. Replace the old filter for new filters. This is done along with changing the chemistries to ensure a clean, uncontaminated system.
4. Mix developer replenisher according to manufacturer's instructions. Fill the replenisher tank, including developer starter.	4.1. In order to maintain proper quality control, chemistries must be mixed correctly.

5. Mix fixer and fill fixer replenisher tank.
6. Turn on the replenisher pumps and continue to operate them until the developer enters the processor tanks.
7. Again drain the processor tanks and refill them from the replenishment tanks. This process is known as "seasoning the tanks."
   7.1. This process removes any water that might have been in the lines from the replenishment tanks to the processor tanks.
8. Adjust the developer and fixer pumps to replenish 65 ml of each chemical every 5 minutes.

PROBLEMS AND SOLUTIONS ENCOUNTERED IN FLOOD REPLENISHMENT

Problem: Cost containment of flood replenishment.
Solution: The total volume of fluid will be replenished every 16 hours, resulting in high-quality radiographs.

Problem: Unseasoned developer replenisher.
Solution: Due to the rapid turnover in chemicals, there is *no* chance for the chemicals to deteriorate or oxidize, resulting in a very stable process.

Problem: A different brand of timer than mentioned in the equipment section.
Solution: Set the developer replenisher rate at 780 ml/hour to the processing tank. For other questions on the timer, request technical assistance from the manufacturer of the timer.

Problem: Amount of chemicals to be mixed at one time.
Solution: Do not mix any more chemicals than the amount to be used in a 2-week period of time.

ACCEPTABLE RADIOGRAPHIC PROCESSOR LIMITS

The processor limits are the same for a regular radiographic processor. Base plus fog ±0.05 and ±0.10 in density for the medium density and density difference.

CORRECTIVE ACTION

The corrective action required for quality control data points that are out of the control limits is the same as that for a regular radiographic processor.

Responsibility	*Action*
1. X-ray service engineer or electrician	1.1. To install the replenishment timer for the flood replenishment processor.
2. Quality control coordinator.	2.1. To perform routine quality control checks on the flood replenishment processor.

APPROVED: ______________________

DATE: ______________________

RADIOGRAPHIC PROCESSOR MAINTENANCE PROCEDURE

Overview

Preventive and corrective maintenance are essential parts of quality control and the quality assurance program.

OBJECTIVE

To ensure that the radiographic processors are maintained in a clean and functional condition as per the manufacturer's specifications.

PLAN: GATHER EQUIPMENT

1. Filters
2. System cleaner (this is not used for every cleaning procedure)
3. Extra rollers
4. Extra belts
5. Miscellaneous equipment (there may be specific equipment requirements for the processor as stated in the manufacturer's specifications)

Implementation	*Rationale and Key Points*
1. Establish and maintain processor maintenance log for each radiographic processor.	1.1. Processor log must include cleaning dates, who did the work, preventive and/or corrective maintenance.
2. Immediately after shutting a processor down, wipe all surfaces with a damp cloth to assure that the processor is wiped clean of all chemicals.	2.1. Chemicals that are left on the processor are potential sources of contamination and corrosive to the equipment.
3. Clean crossover racks daily.	
4. Processor racks should be pulled, checked, and cleaned weekly. Flush-	4.1. A clean processor functions well and requires little maintenance.

Implementation	*Rationale and Key Points*
ing with water for this process is sufficient. Do not use systems cleaner each time the processor is cleaned.	*4.2.* System cleaner by nature is a very strong cleaning solution. It can be corrosive if used too frequently.
5. Always complete a sensitometric check after cleaning, preventive maintenance or corrective maintenance.	*5.1.* Ensure that the processor remains within the established control limits.
6. If processor data points are out of the control limits, first THINK, then begin the troubleshooting process systematically.	*6.1.* Do not change more than one processor parameter at a time before rechecking the processor with a sensitometric control strip.
a. Check the normal operating levels.	
b. Check the maintenance log and continue from there.	
7. **Stop, look, and listen.** The processor is like a car. It has its own hum when it is working properly. Be alert for trouble if the processor begins to sound differently.	
8. Use of system cleaners:	
a. **Safety first.** Always use protective glasses or goggles when working with these chemicals.	*8.1.* System cleaning chemicals are caustic.
	8.2. The system chemicals will corrode metal to which they are exposed.
b. **Never** submerge the racks into the cleaner.	*8.3.* These chemicals will react with one another. If the chemicals are placed in the wrong tank, a chemical reaction will occur.
c. Use the system cleaners separately. Never mix them and do not use them in the wrong tank. **Haste makes waste.**	
d. Flush the tanks well with water after cleaning. Run the recirculation pumps.	*8.4.* This process will clean the tanks of the chemicals.
e. Season the processor after cleaning for 15 to 20 minutes. **Do not** process any films. Seasoning is done by filling the developer and fixer tanks with chemicals and letting them sit.	*8.5.* Seasoning of the tanks aids in removing residual system cleaner from the processor tanks.
	8.6. By adding new chemistry from the replenishment tanks, fresh, non-contaminated chemicals are assured.
f. Now empty the tanks of the chemicals and refill with fresh chemicals from the replenishment tanks.	*8.7.* The processor is now ready for sensitometric control strips to be run and quality control to be re-established.

g. Document the use of systems cleaner, date, person who cleaned system, and data from quality control run on the processor maintenance log and the control charts.

Responsibility	*Action*
1. Quality service engineer	*1.1.* Use of systems cleaner to clean processor.
2. Quality control coordinator	*2.1.* Performs quality control testing on the cleaned processor and ensures that the processor is within the control limits.

APPROVED: ______________________

DATE: ______________________

PROCESSOR MAINTENANCE LOG

Date	Cleaning	Preventive Maintenance	Corrective Maintenance	System Cleaner	Name

PROCESSOR MAINTENANCE LOG

____

Processor #______________________

Developer temperature: ______________

Developer temperature calibrated: ______________

Developer replenishment rate: ______________

Fixer temperature: ______________

Fixer replenishment rate: ______________

Water temperature: ______________

Water temperature calibrated: ______________

Water flow rate: ______________

Chemicals changed: ______________

____

____

Developer tanks filled: ______________

____

____

____

Developer filter changed: ______________

____

____

____

Fixer replenisher tank filled: ______________

____

____

____

Water filter changed: ______________

____

____

(continued)

Racks cleaned: ______________________________

____

____

____

____

Comments: ______________________________

____

____

____

____

____

____

____

____

RADIOGRAPHIC PROCESSOR CHECKLIST

1. Warm-up time of 20 minutes for equilibration before starting quality control procedures.
2. Temperature control:
   a. Developer
   b. Water
   c. Dryer
3. Check and change filters as needed.
4. Check replenishment rates.
5. Check water flow rates.
6. Observe chemical mixing.
7. Check to see that the densitometer is zeroed and calibrated.

RADIOGRAPHIC PROCESSOR TROUBLESHOOTING CHECKLIST

Overview

There are a few rules that never should be broken when troubleshooting a piece of equipment or a procedure. The first of these rules is to think before you take action. A small graphical error may send you on a very misleading troubleshooting excursion. Verify your readings by redoing a quality control run. The second rule is do not use the shotgun approach to troubleshooting. The shotgun is avoided by changing only one parameter at a time and then rechecking quality control.

TROUBLE SHOOTING CHECKLIST

1. Exceeded control limits:
   a. Temperature of the developer
   b. Temperature of the water
   c. Temperature of the dryer
   d. Replenishment rate
   e. Water flow rate
   f. Recirculation of chemicals
   g. Filters: developer, water, and air
   h. Chemical solution mix date
   i. Preventive and/or corrective maintenance
   j. Film fog
   k. Control emulsion
   l. Transport time
2. Trends and drifts in the control chart:
   a. Temperature of the developer
   b. Replenishment rate
   c. Leaks or overflow from the fixer tank to the developer tank (resulting in contamination)
   d. Chemical mixing

   e. Little gremlins
   f. Control emulsion age or fog
   g. Change in the chemical mix, brand, or film
3. Words of wisdom:
   a. Change only one adjustment or parameter at a time. Repeat the sensitometric strips.
   b. Shotgun approach of changing many parameters simultaneously is a nonproductive, time-wasting, and costly approach to troubleshooting any problem.
   c. Once a change has been made in one or more parameters, document it. Include why, what, how much, outcome, date, and name in documentation.

SILVER RECLAMATION: EVERY DARKROOM A POTENTIAL SILVER MINE

Overview

Every radiographic department, no matter how small, should have a program of silver reclamation. Usually the money received from such a program will **pay** for the **entire** amount of fixer used in the department on a yearly basis.

EQUIPMENT

Silver reclamation equipment is available from radiographic processing and film companies. This equipment usually consists of a silver reclamation unit for the processor. Test papers and solutions are available from these manufacturers to assess the amount of silver being retrieved from the spent radiographic fixer. This equipment must be monitored with some type of a quality control program to assure maximum benefit for the radiology department.

Implementation	*Rationale and key points*
1. Proper installation of the silver reclamation unit.	*1.1.* This unit should be installed in accordance with the manufacturer's instructions.
2. This unit must be tested periodically to ensure the adequate reclamation of silver from the spent fixer.	2.1. These tests are completed by using test papers and solutions. *2.2.* If the test papers indicate the maximum amount of silver is not being reclaimed, a second or third unit may be connected to the overflow from the first unit.
3. The unit will need to be emptied at specified intervals and the silver turned in for reprocessing.	*3.1.* X-ray service companies often will contract from the hospitals for the silver reclamation.

CINE-RADIOGRAPHY: ESTABLISHMENT OF PROCESSOR CONTROL LEVELS

Overview

The first step in cineradiography processor quality control is the establishment of the processor control levels.

PLAN: GATHER EQUIPMENT

1. Master control monitoring film
2. Sensitometer
3. Densitometer
4. Ruler, scissors, pencil, and the like
5. Control charts

Implementation	*Rationale and key points*
1. Turn on processor: a. Set the developer, water, and dryer temperature. b. Adjust the replenishment rate of the processor to the level specified for the amount of film to be processed. c. Allow the processor a 20-minute warm-up period.	1.1. Ensure that the manufacturer's instructions for film conditions are strictly followed. 1.2. All electronic equipment needs a warm-up period for equilibration.
2. Select one roll of film as the master control film. a. Open roll of film in darkroom in complete darkness. Remove from plastic bag. b. Cut a control strip about 12 inches long. c. Return film to bag and can; store for future use.	2.1. The master film will be used for the control film for the entire lot number of cine film.

Implementation	*Rationale and key points*
3. Expose the film (control strip) to a sensitometer. a. Insert the control strip into the sensitometer with the emulsion side down. b. Push the exposure switch. c. Remove the control strip from the sensitometer once it has been exposed.	3.1. The sensitometer places a reproducible known exposure through a step wedge on the film.
4. Processing the control strip. a. Attach the control strip to a leader with waterproof tape. b. Insert the control strip into the cine processing box (emulsion side down).	
5. Process a total of four control strips: one each hour for 4 hours.	5.1. At a minimum, four control strips need to be run to obtain control limits.
6. Calibrate the densitometer with the manufacturer's reference strip.	6.1. The quality control equipment must also be calibrated to maintain accuracy.
7. Measure and record the densities in zones 1 through 4 for each of the cine control strips. a. Zone 1: unexposed portion of the film. Base plus fog. b. Zone 2: density closest to 0.50. c. Zone 3: density closest to 1.40. d. Zone 4: most dense portion of film.	7.1. Documentation is an important step in the quality control process.
8. Calculation of the control level aim points: a. Determine the average density values for each density zone by averaging the four values together for each zone.	8.1. By running four control strips and averaging the results, the average density of each step over a specified amount of time (4 hours) is determined.
9. Documentation: a. Enter the numerical data thus determined as the control level aim points on the control chart. b. Record the maximum and minimum density values in the space provided. c. Record the values obtained from zones 2 and 3 on the control chart in their designated places.	9.1. The values that are now recorded are considered to be the true limits for this particular batch of emulsion film.

Responsibilities	Action
1. Quality control coordinator or designee	1.1. Establishes the control for each new film batch.

APPROVED: ____________________

DATE: ____________________

Cine Film Processor

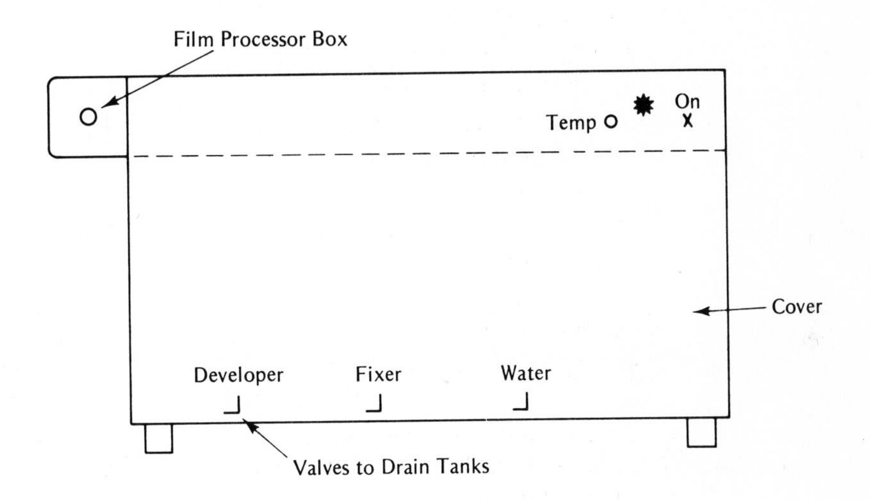

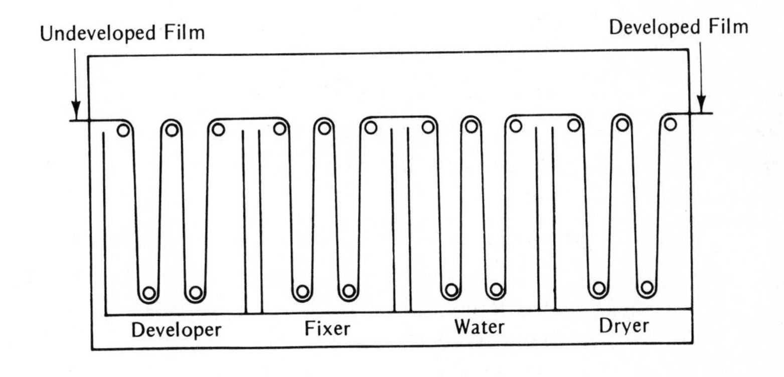

DAILY CINE PROCESSOR QUALITY CONTROL

Overview

There is no rigidity in the number of control strips that are run during any time interval to check cine or radiographic processing. However, there is a correlation between processor quality control and optimum radiograph quality. The minimum number of control monitoring strips that can be done in a 24-hour period for quality control is one.

With the start-up of a new cine processing quality assurance program, it is desirable to process several control strips for the first few weeks. This will allow the quality control coordinator to become aware of random variation, drifts, and trends. It also enables the quality control coordinator to become familiar with film quality, replenishment rates, and patient load.

As the cineradiography laboratory settles into a set pattern, and the quality control coordinator becomes more familiar with his or her equipment, the number of control monitoring strips will be decreased. At this point, one control strip per roll of film is usually sufficient.

PLAN: GATHER EQUIPMENT

1. Master control film
2. Patient film
3. Sensitometer
4. Densitometer
5. Control chart
6. Ruler, scissors, waterproof tape, pencil, and the like

Implementation	*Rationale and key points*
1. Prepare and process one strip from master control film. (Use the same procedure used for establishing control limits.)	1.1. This is a quality control check on the processor compared with the following day.

Implementation	*Rationale and key points*
2. Process one cine control monitoring strip from the patient film.	2.1. This is a quick and efficient check to determine if a problem with the individual roll of film exists. It is also another processor check.
3. Calibrate the densitometer.	3.1. The calibration of the densitometer with the manufacturer's reference strip is as important as the reliability of the instrument itself.
	3.2. A small error in the calibration of the densitometer could lead to misinterpreted control strips.
4. Measure the densities of zones 1 to 4.	
5. Document and plot the density data points on the control chart.	5.1. Documentation is one of the keys to quality control and the quality assurance program. It allows for easy interpretation and visualization of the process being monitored.
6. Evaluate the cine processor according to the aim points. If the data points are out of control, refer to the corrective action section.	6.1. A quick look at the control chart will tell you if a drift or trend is beginning to appear, as well as random variation.

CORRECTIVE ACTION

1. Reasons for corrective action:
   a. Data points outside control limits.
   b. Three to five data points headed in the same direction (trend, drift).
   c. Three to five data points are consistently above or below the aim points.
2. Implementation for corrective action:
   a. Recalibrate the densitometer.
   b. Reread the control strip to eliminate a reading error.
   c. Examine the processor parameters.
      (1) Temperature of developer, water, dryer
      (2) Replenishment rate
      (3) Transport speed
      (4) Agitation
   d. Redo the control monitoring strip.
   e. If the control monitoring strip remains out of control limits, institute the troubleshooting procedure for cine-radiography.

Responsibility	*Action*
1. Quality control coordinator or designee	*1.1.* Completes daily quality control and documents the results.
2. Chief technologist	*2.1.* Assists in the process of trouble-shooting quality control problems when the need arises.

APPROVED: ______________________

DATE: ______________________

CINE-RADIOGRAPHIC PROCESSOR CHECKLIST

1. Warm the processors up for at least 20 minutes for equilibration before starting quality control testing.
2. Temperature checks:
   a. Developer
   b. Water
   c. Dryer
3. Chemical mixing: assure that the replenisher has a floating lid with a dust cover.
4. Recirculation of chemicals: potential problems:
   a. Improper agitation
   b. Poor replenisher mixing
   c. Poor temperature control
5. Check the water flow rate and pressure.
6. Check the transport time by processing a known amount of film with a stopwatch.
7. Check to see that the densitometer is zeroed and calibrated.

CINE TROUBLESHOOTING PROCESSOR CHECKLIST

Image Quality	*Control Strip*	*Causes*	*Corrective Action*
1. Density too low	Normal range	Underexposure	1. Technolgist error 2. Readjust exposure F-stop
2. Density too high	Normal range	Overexposure	1. Technologist error 2. Readjust exposure F-stop
3. Density too high	Zone 1 density too high	Light leak Safe light	1. Check darkroom integrity 2. Technologist error; film must be handled in darkness
4. Density too low	Density too low	Developer temperature too low	1. Check temperature
		Processor speed too fast	2. Check speed
		Replenishment too slow	3. Increase replenishment rate
		Contaminated developer	4. Remix solution
		Improperly mixed solutions	5. Remix solution
		Inadequate developer circulation	6. Check filter and pump system
5. Density too high	Density too high	Developer temperature too high	1. Check temperature
		Processor speed too low	2. Increase processor speed
		Flood replenishment	3. Decrease replenishment rate

EVALUATION AND ESTABLISHMENT OF NEW CONTROL EMULSION LOT NUMBERS CROSSOVER PROCEDURE

Overview

When the master control film or regular radiograph film lot number is nearly depleted, perform a crossover procedure to introduce a new emulsion lot number. The procedure is comprised of running several pairs of control sensitometric monitoring strips from both the old and the new lot numbers simultaneously. This procedure is done to determine if there are any density differences between the two lot numbers of film.

During this procedure any necessary chemical and/or mechanical adjustments to the processor should be based solely on the densities of the old master control monitoring strips.

With the processing control still based on the old control strips, the density parameters for the new control emulsion lot number are determined in the usual manner.

OBJECTIVES

1. To assess the density of a new emulsion batch lot number of cut film, roll film, or 35, 70, or 105 mm film.
2. To compare the densities of the new film to those of the old film through sensitometric and densitometric readings.

PLAN: GATHER EQUIPMENT

1. Old control film
2. New control film
3. Sensitometer
4. Densitometer
5. Control charts specific for crossover procedure
6. Ruler, scissors, pencils, and the like

Implementation	*Rationale and key points*
1. Ensure that all the processors are in control with the old emulsion batch.	1.1. This excludes the problem of introducing another variable when doing the crossover test.

Implementation	*Rationale and key points*
2. Simultaneously process three to four controls trips from the old and new emulsion batches in each processor.	2.1. This will allow for the establishment of new density levels for the new emulsion batch.
3. Read, document, and average the B + F, MD, and DD for the three or four strips of the old and new emulsion batches.	3.1. This is a method to determine if there are density differences between the new and old emulsion batches.
4. Subtract the average old emulsion values from the new emulsion values to obtain the density difference.	4.1. The density differences are normally very small.
5. Take the average of the differences for all three photographic points for the processors.	
6. Add the average density differences to the old operating levels to determine the new operating level.	6.1. The old operating level was used in calculating the densities of the new emulsion batch.
7. Record these new operating levels on the control chart and use these levels for daily quality processor control.	

EXAMPLE OF TABLE USED IN TABULATING THE NEW OPERATING LEVELS FOR A NEW EMULSION BATCH

	B + F	*MD*	*DD*
Processor A			
Average new emulsion	0.18	1.18	1.90
Average old emulsion	−0.20	−1.17	−1.87
Difference	0.02	0.01	0.03
Processor B			
Average new emulsion	0.22	1.24	1.94
Average old emulsion	−0.20	−1.26	−1.89
Difference	0.02	−0.02	0.05
Processor C			
Average new emulsion	0.16	1.14	1.89
Average old emulsion	−0.20	−1.12	−1.87
Difference	−0.24	0.02	0.02
Average of difference	0.00	0.01	0.03
Old operating levels	0.24	1.14	1.90
New operating levels	0.24	1.15	1.93

Responsibility	*Action*
1. Quality control coordinator	1.1. Establishes new operating levels for new emulsion batches.

APPROVED: ____________________

DATE: ____________________

DARKROOM INTEGRITY, SAFE LIGHTS, FILM BINS, AND PASS BOX QUALITY CONTROL PROCEDURES

Overview

A darkroom is an integral step in maintaining the high quality of the radiograph after it has been exposed to radiation. A light leak at this point could fog and thus greatly diminish the overall physical quality of the radiograph.

OBJECTIVES

1. To ensure that the darkroom is safe from all unwanted sources of light or radiation that may fog the radiograph.
2. To ensure that the safe lights are in proper working order.
3. To ensure that the film bins are light tight.

PLAN: GATHER EQUIPMENT

1. Unopened box of radiographic film
2. Aluminum step wedge, attenuator
3. An acceptable film-screen cassette (one that has been through quality control checks)
4. Black paper (one-half the size of the film to be checked)
5. Densitometer
6. Clock with second hand or stopwatch

Implementation	*Rationale and key points*
1. Choose the first darkroom to be tested for quality control.	1.1. The frequency of quality control checks should be every 4 to 6 months.
2. Enter the darkroom and turn off **all** lights (include the safe lights).	2.1. Allow 15 to 20 seconds for your eyes to adjust to the dark.
3. Carefully examine the darkroom for visible light leaks. a. Cracks around doors	3.1. These are the most common areas for light leaks to occur.

Implementation	*Rationale and key points*
b. Cracks around the opening of the processor *c.* Cracks around the pass box *d.* Extraneous light from the ceiling	
4. If a source of light is found, record it and take appropriate action.	*4.1.* A problem at this point would negate continuing with the quality control checks.
5. Cover any indicator light with opaque material and proceed to open the film box. The next checks need to be done for each type of film to be used in that darkroom.	*5.1.* Even one small indicator light on a replenishment pump or processor may fog the film.
6. Load the film into the cassette in total darkness. Do not use the safe lights.	*6.1.* Prior to starting the procedure, the cassette must have been checked for integrity.
7. Radiograph the step wedge with it centered on the cassette. A sensitometer may be used for films other than general radiographic films with an exposure of 1.0 density.	
8. Return to the darkroom, cover the indicator lights, and open the cassette. Cover one-half of the film with the black paper. This will cover one-half of the latent image.	
9. Close the cassette and turn on the safe lights and expose the film to them for 1 minute.	*9.1.* This portion of the test is to see if the safe lights are functioning properly.
10. Remove film from the cassette and process at once.	*10.1.* A delay in processing the film may add new variables to the test.
11. Repeat steps 6 to 10 and expose the indicator lights as well as the safe lights. Expose the film for 2 minutes.	
12. Process the films.	
13. With the densitometer, measure: a. Base plus fog of the film not exposed to the safe light. Closest to 1.0 density. *b.* Measure area of the film that was exposed to the safe light. c. Measure area of the film that was exposed to the safe light plus the indicator light.	*13.1.* The acceptable limits in density from the side of the film not exposed to the safe lights to the side of the film that was exposed to the safe lights should not exceed 0.02 for 1-minute exposure. *13.2.* The 2-minute exposure to safe lights plus the indicator light should not exceed 0.05.
14. Record the results of these densities on quality control charts.	

Implementation	*Rationale and key points*
15. After the integrity of the darkroom has been determined, the film bin and the pass boxes need to be checked in a similar manner for light leaks. **Film bin:** Remove a film from the film bin and process. Record the density and observe for fogging. **Pass box:** With the integrity of the darkroom and the cassette established, if a film (exposed or unexposed) demonstrates fogging, the problem has been found by the process of elimination.	

AVOIDING PROBLEMS

1. Use tested cassettes for your quality checks.
2. Use the same film as is used in that darkroom.
3. If other than general radiographic films are used in the darkroom, use a sensitometer with an exposure for 1.0 density to expose the film.
4. Calibrate the sensitometer and densitometer before use.
5. The densities of the films must be read on adjacent areas to avoid exposure variations as well as processor variations.
6. Check each radiographic darkroom in **exactly** the same manner.

TROUBLESHOOTING PROBLEMS

Problem: Density is greater than the established acceptable limits.

Solutions: Is the safe light filter present?

If the filter shows signs of fading, cracking, or other wear, replace it.

Check the wattage of the bulb in the safe light.

What is the distance from the safe light to the working area? These specifications may differ slightly from one manufacturer to another.

Indicator lights may have to be covered if this portion of the film is where the problem exists.

Responsibility	*Action*
1. Quality control coordinator	1.1. To ensure that the darkroom, pass boxes, and safe lights are checked by the standards established for the radiographic facility.

APPROVED: ____________________

DATE: ____________________

MAINTENANCE PROCEDURES FOR VIEW BOXES, HOT LIGHTS, AND PROJECTORS

Overview

These are areas often forgotten in quality control and the quality assurance program. If the actual viewing systems are not up to par, the physicians may be unhappy regardless of the physical quality of the radiograph.

1. **View boxes:**
   a. What is the surface condition of the view box?
      Scratches can interfere with the reading of a radiograph.
   b. Is the light source consistent with time?
      Older view boxes do not have good luminescence and thus detail is difficult to identify on a radiograph.
   c. Are the view boxes similar throughout the department?
      This may be a contributing factor to discontent in the radiographic department. Radiographs viewed on one system may differ in physical quality when viewed on another system. Therefore, consistency is important throughout the department.
   d. View boxes are similar to any other piece of radiographic equipment. They must be maintained in good working order and cleaned on a regular basis. This inudes the outside (surface) and the lamps.
2. **Hot lights:** The same parameters that pertain to the view boxes pertain to the hot lights. Cleaning and maintenance are important and should be done on a regular basis.
3. **Projectors:** The same parameters that pertain to the view boxes pertain to the projectors. The cleaning of the film gates and the lenses are important factors in maintaining good visual acuity.

ASSESSING TECHNICAL ERROR

Overview

Technical error in a radiology department can be very costly in time lost (poor room utilization, poor personnel utilization), expense of supplies (repeat films), and a greater amount of radiation dose for the patient. The latter cannot be measured in dollars and cents.

OBJECTIVES

1. To reduce the number of repeat films by appropriate in-service education for all employees.
2. To readily identify technical problems that may need corrective action.

PLAN

1. To set established operating standards.
2. To acknowledge the six most common problems due to technical or technologist error.
   a. Overexposed radiographs
   b. Underexposed radiographs
   c. Patient motion
   d. Incorrect positioning of the patient
   e. Incorrect number or type of views taken
3. To find a means of correcting these identified problems.

Implementation	*Rationale and key points*
1. Provide each technologist with the abbreviations used for examinations and views in the department.	1.1. Promotes organization and effective communication throughout the department.
2. Provide each technologist with a list of required views for each examination. The radiologist may write these for you.	2.1. Will decrease the number of times that the patient will have to return to x-ray for repeat films.

Implementation	*Rationale and key points*
3. In plastic protective covers, post current technique charts in each room. They must be legible.	3.1. This will give the inexperienced technologist a starting place. It also promotes reproducibility.
4. Standardize kVp, mAs, and mA as much as possible within the department.	4.1. May allow similar techniques to be used in similar rooms.
5. Record exposure factors including fluoroscopy time on the x-ray request form and in the patient's chart.	5.1. This has become a safety rule that is endorsed on the state and federal level by HEW, Radiology Division.
6. Standardization of films, solutions, and screens.	6.1. This will decrease the number of variables that may interfere in any given process.
7. Make adequate equipment available for patient measurements. a. Calipers b. Angulators c. Tape measures d. Miscellaneous	
8. Have safety straps and immobilization equipment available in every room.	8.1. It is a poor and dangerous practice to leave a patient to gather needed equipment (in a court of law, this constitutes abandonment).
9. Promote and encourage the technologist to THINK before he or she fires off an exposure.	
10. Make the NCRP report no. 33 available to all x-ray technologists.	

Responsibility	*Action*
1. In-service education coordinator	1.1. Assesses the technologists' beginning abilities and guides them toward expertise.
2. Quality control coordinator	2.1. Ensures that the equipment is working properly and within the established operating standards for the department.

APPROVED: ____________________

DATE: ____________________

RADIOGRAPHIC FORM: EVALUATION OF NEW PRODUCTS

1. Name of product: ______________________________
2. Use of product: ______________________________
3. Company who produces product: ______________________________

   ____
4. Department requesting product: ______________________________
5. Other departments where product may be used:

   ____
6. Length of test period: ______________________________
7. New item: ______________ Replace old item: ______________
8. Cost of item: ______________________________
9. How is item packaged? ______________________________
10. Will company supply item at their expense for the test period?

    ____
11. If appropriate, does the item have state and federal approval?

    ____
12. Will item be used in direct or indirect patient care?

    ____
13. If item is electrical, it must be inspected for safety by the hospital clinical department.

Requested by ______________

Department head ______________

Approved for test period ______________

EVALUATION REPORT

1. Did the product meet its specifications?

2. How will the product benefit your department? hospital?

3. Were thee areas that did not meet your requirements?

4. Will this product promote cost containment?

5. Statement from clinical engineering or other related areas.

Approval:	______________ Department head	Disapproval:	______________ Department head
	______________ Administrator		______________ Administrator

PURCHASE SPECIFICATIONS FOR X-RAY EQUIPMENT

Overview

The initial step in acquiring new equipment is to determine the desired performance specifications. Specifications may be stated in terms of the desired performance level. The manufacturers (vendors) may be informed of the functions that the radiographic facility wishes the equipment to perform.

The sales representative of the manufacturer or vendor may, through negotiations with the radiographic facility, trade off or balance out the cost of the equipment versus the need for the specified performance requirements of the equipment.

FINAL EQUIPMENT QUOTATION

The final equipment quotation will include:

1. Purchase specifications
2. Performance specifications
3. Agreement or disagreement to contract terms
4. Delivery time
5. Installation time
6. Warranty information
7. Service information
8. A general statement that addresses the topic of state and federal laws

PURCHASE SPECIFICATIONS

1. All equipment and related components therein must meet HEW regulations.
2. All the individual components must be compatible with the x-ray system, resulting in a functioning unit. The unit must meet purchase and performance specifications.
3. If deviations from specifications (either better or less specific) are necessary, they may only be included by written permission of the purchaser of the equipment.
4. The manufacturer or vendor must supply:
   a. Complete set of equipment specifications

b. Complete set of schematics (blueprints) of the equipment
c. One or two sets of operating manuals
d. One or two sets of service manuals
e. In-service education for the technologists

Equipment *Specifications*

I. **Generator**
   A. Single or three-phase
   B. Voltage supply
   C. Kilovoltage (kV):
      1. Number of pulses
         a. 1
         b. 2
         c. 6
         d. 12
         e. Constant potential
      2. Percent of decrease in rating with falling load
      3. Line voltage regulation manual, automatic
      4. Minimum kV
      5. Maximum kV
      6. kV rating
      7. Accuracy, precision, kV
      8. kV increments
      9. Rectification:
         a. Silcon
         b. Valve tube
         c. Thermonic
      10. Ripple load factor
   D. Milliamperage (mA):
      1. Minimum settings
         a. Small focal spot
         b. Large focal spot
      2. Maximum settings
      3. Accuracy or precision
      4. mA increments
         a. Small focal spot
         b. Large focal spot

Equipment	Specifications	
E. Timing controls:		
1. Minimum setting		
2. Maximum setting		
3. Timer accuracy		
4. Time selector increments		
5. Timing control synchronous or nonsynchronous		
6. Exposure termination time		
7. Decimal or fractional time display		
8. Type of contacting mechanical: solid state		
II. Image intensifier:		
A. Input phosphor material:		
1. Cesium iodide		
2. Cadmium		
B. Size of output phosphor		
C. Usable diameters of input phosphor		
D. Line pair resolution		
E. Contrast ratio	Candela/cm	
F. Conversion ratio	Millirad/cm	
G. Image intensifier tube gain		
H. Guaranteed useful life	hours/months	
III. X-ray tubes:	Radiographic	Fluoroscopic
A. Number of tubes		
B. Maximum kV rating		
C. kW ratings		
D. Targaret diameter		
E. Anode heat storage capacity		
F. Housing heat storage capacity		
G. Anode rotation speed		
H. Focal spot size		
1. Small		
2. Large		
3. Biased		
I. Heat dissipators		
J. Targaret angle		
K. Anode heat monitor		

Equipment *Specifications*

IV. Automatic brightness control:

A. Fluoroscopy
  1. Photomultiplier tube
  2. TV
  3. Controlled mA, and/or kV

B. Fluoroscopic camera
  1. Photomultiplier tube
  2. Ion chamber
  3. Controlled mA, kV, or exposure length

V. Automatic exposure termination:

A. Yes–No

B. Photomultiplier tube or ionization chamber

C. Density control

D. Response time ± 2 ms

E. Tracking accuracy versus kV

F. Tracking accuracy versus mA

G. Forced exposure termination: yes? no?
Response time

H. Phototimer cassettes required: yes? no?

VI. Viewing mechanics:

A. Mirror–optics: yes? no?

B. Television chain: yes? no?
If yes
  1. Optical connecting system:
    a. Fiberoptics
    b. Tandem lens system
  2. Television camera:
    a. Line rate
    b. Vidicon or plumbicon standard—medical
    c. Vidicon or plumbicon select–medical
    d. Band width
  3. Television monitor:
    a. Contrast ratio
    b. Line rate
    c. Band width
    d. Size:
      (1) Diagonal
      (2) Length

Equipment	*Specifications*

VII. Recording media:

A. Fluoroscopic cameras:

1. Cine

a. 35 mm or 16–16 super

b. Frame rate:

(1) 30 per second

(2) 60 per second

(3) 90 per second

(4) Other

c. Overframing percentage

2. Spot film camera:

a. 70 mm

b. 90 mm

c. 100 mm

d. 105 mm

B. Spot film:

1. Yes? No?

2. Front loading

3. Rear loading

4. Cassette size:

a. 8 × 10

b. 9½ × 9½

c. Grid

5. Formats, variables 1/1, 2/1, 4/,1, etc.

C. Video tape recorder:

1. Reel to reel: tape size

2. Video cassette

3. Video disc recorder

D. Computer system:

1. Digital intravenous arteriography system

VIII. Tube hangers:

A. Detents: mechanical or electrical

B. Floor to ceiling mounted or ceiling suspended

C. Measurement accuracy

D. Tube rotation

E. Minimum source image distance

IX. Collimators:

A. Types:

1. Rectangular or iris

Equipment	*Specifications*

2. Combination field

B. Alignment

C. Source-image distance indicator

D. Aluminum filtration equivalency

E. Added filtration

F. Tube angulation indicator

G. Slots for wedge filters

X. Radiographic–fluoroscopic table:

A. Types:

1. Stationary
2. Tilting
   a. Vertical to horizontal
   b. Horizontal to trendelenberg

B. Table tops:

1. Flat
2. Curved
3. Cradle
4. Movable, two-way, four-way
5. Power or float top

C. Automatic centering:

1. Cross table of table top
2. Longitudinal of table top
3. Spot film device
4. Over table tube

D. Automatic stop at horizontal:
Yes? No?

E. Bucky grid

F. Bucky speed:

1. High
2. Standard

G. Table top to image receptor distance

XI. Auxiliary equipment:

A. Digital fluoroscopy system

B. Tube overload indicator

C. X-ray exposure counter

D. Automatic high speed rotation control

XII. Delivery date:

A. 90 to 180 days after agreement

Equipment *Specifications*

B. Dependent upon approval of certificate of need

C. Contingency plans

D. Method of payment:

1. 10% down payment

2. 70% arrival of equipment

3. 20% turn over of equipment

XIII. Installation of equipment:

A. Agreement

B. Insurance

C. Penalty clauses

XIV. Performance requirements:

A. Tests to assure that the equipment meets specifications

B. Performance guaranteed by bond or other means

XV. Acceptance of equipment:

A. All requirements for installation and performance meet the standards of the hospital

XVI. Warranty considerations:

A. Length of service

B. Component parts guaranteed

C. Prorated warranties

EXAMPLE OF X-RAY EQUIPMENT SPECIFICATIONS FOR AN X-RAY SPECIAL PROCEDURE ROOM

The specifications listed below are target specifications. Any deviation from them must be listed and clarified.

I. Biplane System:

A. Ap-C-Arm or equivalent (mechanical):

1. Cranio-caudal: 45 degree rotation
2. Caudo-cranial: 45 degree rotation
3. Oblique angulation: 90 degree rotation
4. a. What is your maximum distance in inches utilizing the Sedlinger (femoral) technique of imaging from the femoral area to the apex of the heart?

   ____

   b. How is this movement obtained?

   ____

   ____

   ____

5. a. May the x-ray tube on the C-Arm be used for cut film?
   Yes ________ No ________

   b. Ap position ______________________

   Lateral position ______________________

6. a. Does the C-Arm have an "out of the way" or park position for easy access to the patient?

   ____

   b. How is this position obtained?

   ____

7. Where are the controls for movement of the C-Arm mounted?

   ____

B. Lateral C-Arm or equivalent (mechanical):

1. Cranio-caudal: 45 degree rotation
2. Caudo-cranial: 45 degree rotation
3. Oblique angulation: 90 degree rotation

4. If the Ap system fails, are you able to move the lateral into the Ap position?
   Yes ____________ No ____________
5. a. Can you use the lateral x-ray tube for cut film work?
      Yes ____________ No ____________
   b. Ap position ____________
      Lateral position ____________
6. Where are the controls for the C-Arm mounted?
   ____
7. a. Does the lateral tube have a park position?
      Yes ____________ No ____________
   b. How is this position obtained?
      ____

C. Ap and lateral C-Arms (mechanical):
   1. Is there an isocenter? Yes ____________ No ____________
   2. Do the Ap and lateral arms of the x-ray system track automatically while panning?
      Yes ____________ No ____________
   3. Can both of the C-Arms move independently of one another?
      Yes ____________ No ____________

D. Ap and lateral C-Arms (imaging):
   1. Image intensifier:
      a. 9", 6", and 4" modes
      b. Statement of resolution for each mode. Include the number of line pairs per mm on system 35 mm film
         ____
         ____
         ____
         ____
      c. What is your contrast ratio? ____________
      d. Is a selected image tube obtainable? ____________
         If yes, at what price? ____________
   2. IV system: Ap and lateral:
      a. 875 line IV system, high resolution
      b. Plumbicon camera tubes
      c. Video tape recorders: Ap and lateral
         (1) Reel to reel 1" with slow motion, single framing, and stop modes
         (2) Must have automatic start with cine exposure
      d. Disc storage
   3. 35mm camera Ap and lateral:
      a. Arriteckno, 35mm or equivalent
      b. 90 frames per second
      c. Overframing lens
      d. Two 400 foot magazines
      e. Ability to change F-stops

4. 100 or 105 spot film camera, Ap
   a. Cut film? or roll film? In ______________________ ?
   Out ______________________ ?
   b. 6 frames per second.
5. Cut film changer:
   a. 14 × 14 films
   b. 3 frames per second
   c. Grid
   d. Universal stand so it may be used for Ap and lateral

II. X-ray tubes:

A. Ap and lateral tubes must be identical
B. Heat units: warm anode storage 1,000,000 units
C. Heat units: housing, storage must be water or oil cooled to prevent heat buildup.
D. Starting time for the high-speed rotor under 1 second.
E. Can you use fluoroscopy with a stationary anode?
   Yes __________ No __________
F. Focal spot size:
   0.6 mm and 1.2 mm
G. Can 0.6 mm be biased to a smaller size? ______________________
H. What is your availability of a replacement x-ray tube? Please state time and location

____

____

____

III. Generator–Biplane:

A. Generator must be three-phase constant potential
B. Cine mode:
   1. Capability of kV ______________________
   2. Capability of mA ______________________
   3. Capability of pulse width ______________________
C. Automatic Brightness Control (ABC):
   1. kV controlled? ______________________
   2. Response time ______________________
   3. Is the area of the automatic brightness pickup adjustable in size and position?
   Yes __________ No __________
D. Cut film:
   1. Capability of kV ______________________
   2. Capability of mA ______________________
   3. Capability of time ______________________
E. Spot film, 100 or 105 mm:
   1. Capability of kV ______________________
   2. Capability of mA ______________________
   3. Capability of time ______________________

IV. Diagnostic table:

A. Is the table radiolucent from:

1. All planes at heart location? ______
2. Carbon fiber with *no* cradle? ______
3. State the absorption equivalent in aluminum? ______

B. How is the patient secured to the table?

______

______

C. What are the safety features of the table during a cardiac arrest?

______

______

______

D. Are there special neonate or pediatric tops?

______

______

E. What type of table top movement:

1. Float top? ______ Motorized? ______
2. What is the weight of the moving position?

______

______

3. State the dimensions of movement:
   a. Forward ______
   b. Backward ______
   c. Left ______
   d. Right ______

F. Vertical travel: Motorized: ______

Distance? ______

G. Table tilt

1. Trendelenburg ______
2. Reverse trendelenburg ______

H. Are the physiological connections in the base of the table?

______

I. Are there arm rests included? ______
If not are they obtainable? ______

J. Explain the location of the arms during coronary arteriography or angiography utilizing the sones technique.

______

______

______

______

V. Systems:

Fill in the following blanks to complete the statement that your equipment will perform to upon installation.

On ______________________ mode of the image intensifier ______________ line pairs per inch can be seen by the naked eye on 35 mm film with not more than ______________ microRad/frame radiation input to the image intensifier using Emil-Funk resolution phantom and a mid-resolution 35 mm film.

VI. What is your delivery time?

____

VII. What is your installation time?

____

VIII. Make a statement concerning your warranty period.

____

____

____

____

IX. Service:

A. How many service people in this area with special procedure training?

____

B. How many years' training do they have? ______________

C. Availability of service in **time** to ______________ hospital 7 days a week.

X. In-service:

A. Technologist level (mandatory)

B. Two copies of manuals:

1. Service
2. Operational

XI. Please enclose a blueprint of special procedure area:

Bid Biplane C-arm Configuration:

Cost of equipment:

State sales tax:

Delivery and installation:

Total cost of equipment:

RADIOGRAPHIC PROCESSOR SPECIFICATIONS PHYSICIST REQUEST

1. Size and cost
2. Control and indicator sections
3. User education
4. Venting
5. Feed and detection section
6. Water supply system
7. Processing section
8. Washer and dryer sections
9. Recirculation system
10. Replenishment system
11. Sensitometric monitoring
12. Drive train
13. Temperature regulation system

TRAINING MATERIALS AND AIDS

Overview

This section is included for those hospitals that cannot afford formal classroom and laboratory training of their personnel in quality assurance. This may be due to the workload of the department or lack of capital. In this situation, on-the-job training is a suitable alternative. The in-service education director or person directly in control of radiologic continuing education will find that the items listed may serve as supplemental training aids and/or self-instructional training aids for establishing or upgrading a radiologic quality control program.

1. Publications from the United States Department of Health, Education and Welfare.
   DHEW Publications
   U.S. Department of Health, Education and Welfare
   PHS, FDA, Bureau of Radiological Health
   Rockville, MD 20857

   You may write them and request a specific title list relating to Radiologic Quality Control.

2. Resources on quality control and quality assurance programs:
   a. Bureau of Radiological Health
   b. Department of Health and Environmental Control, State level.
   c. Superintendent of Documents, federal level, Washington, D.C.
   d. Film companies:
      (1) Dupont
      (2) Fugi
      (3) IL Ford
      (4) Kodak
      (5) Sakura
      (6) Many others
   e. Manufacturers of quality control equipment:
      (1) Dosimeter Corporation of America
      (2) E.I. duPont de Nemours and Co., Inc.
      (3) Independent X-Ray Dealers Association
      (4) International Radiographic Supply

(5) MDH Industries, Inc.
(6) Nuclear Associates, Inc.
(7) Radiation Measurements, Inc.
(8) Sakura Medical
(9) Supertech, Inc.
(10) Telstar Electronics Corporations
(11) Many others

f. Others (to name a few):
(1) California State Office of Procurement
(2) Dielman Consultants, Inc.
(3) Texas Department of Health Resources
(4) University of Wisconsin

g. Manufacturers of x-ray equipment:
(1) CGR
(2) General Electric
(3) Hewlett-Packard
(4) Phillips
(5) Picker
(6) Seamens
(7) Many other companies

3. Programs available:

a. Title: Bureau of Radiological Health Video Loan Library
(1) Cost: Free on loan
(2) Acquired from: Training Resources Center
DTMA, BRH, FDA
5600 Fishers Lane
Rockville, MD 20852

b. Title: Experiments with X-Rays
(1) Cost: Approximately $225.00 for 15 sets of 10 experiments.
(2) Acquired from: Hewlett-Packard Company
1700 South Baker Street
McMinnville, OR 97128

c. Title: Quality Assurance Testing of Diagnostic Radiographic Units Using an Exposure Meter and Ionization Chamber
(1) Cost: Free
(2) Acquired from: Capintec, Inc.
136 Summit Avenue
Montvale, NJ 07645

d. Title: Radiologic Technology
(1) Cost: Approximately $7.00
(2) Acquired from: Superintendent of Documents
U.S. Government Printing Office
Washington, DC 20402

e. Titles:
(1) Radiographic Density
(2) Principles of X-Ray Protection
(3) Checking the Accuracy of the Light Beam in a Collimator
(4) Calculating Beam Size from a Cone
(5) Operating Principles of X-Ray Machines
(a) Cost: Varies

(b) Acquired from: Nuclear Associates, Inc.
100 Voice Road
Carle Place, NY 11514

(c) Contact: Mr. David Weiner

f. Other programs are available at considerable cost from

(1) Dosimetrics
Education Division
Post Office Box 3244
Cherry Hill, NJ

Quality Control Program approximately $500 per month or $5,000 per year including equipment.

(2) Hewlett-Packard Company
1700 South Baker Street
McMinnville, OR 97128

Cost approximately $4,000.

(3) Other programs from x-ray, film, and equipment manufacturers.

EMERGENCY NUMBERS

Physician Director: Radiology	Beeper Hosp. # Home #
Administrative Director: Radiology	Beeper Hosp. # Home #
Chief Technologist	Beeper Hosp. # Home #
Quality Control Coordinator	Beeper Hosp. # Home #
In-house Service	Beeper Hosp. # Home #
Contract Radiographic Service	Beeper Ans Service # Office # Home #
Angiographic Injector	Beeper Office # Home #
Physiologic Monitoring Equipment	Beeper Office # Home #
Electrocardiographic Equipment	Beeper Office # Home #
Radiographic Processor Equipment	Beeper Office # Home #
Miscellaneous X-ray System	Beeper Office # Home #

APPENDIX: DU PONT TECHNIQUE CHARTS*

TABLE OF CONTENTS

*The material in this appendix is taken with permission from the DuPont Imaging Services Radiographic Technique Guide.

INTRODUCTION

In the formulating of radiographic techniques, it is surprising to find that a complete adult technique chart has not been prepared for distribution. Although many of the technique guides currently available to the profession are excellent, there is a need for a technique chart which is somewhat more complete. However, it must be kept in mind that no technique chart is perfect, but is rather an accurate guide on which the technologist can make changes when indicated. The techniques contained in this booklet are intended solely as a guide which the technologist must adapt to his own practice. Technique changes, when required, can be conveniently made with the help of the Du Pont Bit System of Technique Conversion.

Techniques are based on the premise that the anatomy to be radiographically demonstrated is always measured before consulting the technique chart. Approximately 15% of patients will require technique changes not taken into consideration by the chart. These classifications are:

- Anatomical anomalies
- Disease processes which require an increase or decrease in technique

Additional criteria on which this technique chart is based:

- Time temperature processing
- Matched calibration of all radiographic units
- Single phase generators
- Collimation of the primary beam to the area of interest
- 3 mm aluminum total tube filtration
- Screen exposures: with the following screen type, medical radiographic films:
  CRONEX® 4
  CRONEX 6
  CRONEX 6 Plus

The proposed techniques in this booklet are based on the use of fixed voltages. The kilovolts peak in some examinations may be higher than those currently employed in many radiographic technique charts. With exact collimation of the primary beam, proper filtration, correct grid ratios, and a film with the desired inherent contrast, excellent radiographic quality can be obtained. Some of the advantages of higher kilovoltage techniques are:

- Reduced entrance dosage to the patient
- Shorter exposures which lessen the chance of motion unsharpness
- Increased radiographic latitude
- Improved control of radiographic contrast
- Less heat impressed into the x-ray tube

The unusual feature of this technique chart is that it lists milliampere-second values for both CRONEX Hi-Plus (H-mAs) and Par Speed (P-mAs) intensifying screens. To adapt these techniques to your department, it will be necessary to make a few trial exposures. The following steps are required:

1. Select various portions of the anatomy for demonstration, such as the anterior view of the hand, posterior view of the shoulder, anterior chest, lateral view of the skull, posterior view of the pelvis, and the posterior view of the abdomen.

2. a. Measure the patient and make an exposure with the Guide Milliampere-Seconds (mAs) value indicated.*

b. Make a second exposure which is half of the Guide Milliampere-Seconds.

c. Make a third exposure which is double the value of the Guide Milliampere-Seconds.

3. After the films have been processed, compare them for diagnostic acceptability.

4. If changes are indicated, use the following formula to obtain a correction factor:

$$\frac{\text{Corrected mAs}}{\text{Guide mAs}}$$

The quotient of this problem is then multiplied by the other techniques in its category to maintain a balanced density at all measurements.

5. Example:

Shoulder Technique

cm. Range	H-mAs	
8-11	5	Guide mAs
12-16	6.6	
17-20	10	

On review of the test radiographs, you find that half the Guide Milliampere-Seconds (3.3) produced a film which was underexposed; double the Guide Milliampere-Seconds (13.3) produced a film which was overexposed; the film exposed with the Guide Milliampere-Seconds (6.6) was slightly overexposed. However, an additional film exposed at 5 milliampere-seconds has the desired density scale. See illustrations at right.

Accordingly, the following steps would be indicated:

$$\frac{5}{6.6} = .75$$

The correction factor .75 is then multiplied by all the Guide Milliampere-Second values.
If the exact milliampere-second value cannot be obtained due to limitations in the choice of milliamperage and time, the next closest milliampere-second value may be employed.

cm. Range	H-mAs	
8-11	3.3	Corrected mAs
12-16	5	
17-20	7.5	

*To conserve exposure to ionizing radiation, it is recommended the Guide Film be critiqued before proceeding with step b or c.

A word about technique files. Experience has shown that radiographic techniques should be neatly typed and conveniently displayed. An Acme Visible or Remington Rand Kardex file, holding size 5″ x 8″ cards, would be excellent for this purpose.

The art and science of medical radiography is a demanding skill, and the technologist must constantly strive to improve radiographic quality. We hope these proposed techniques will be of assistance to you.

TERRY R. EASTMAN, R.T.
Imaging Specialist
E. I du Pont de Nemours & Company (Inc.)
Photo Products Department

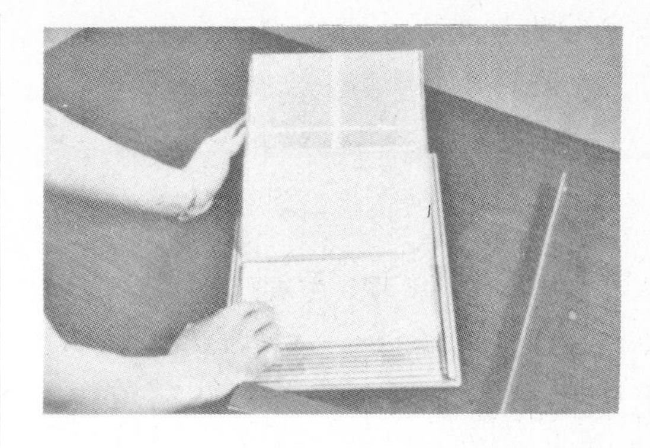

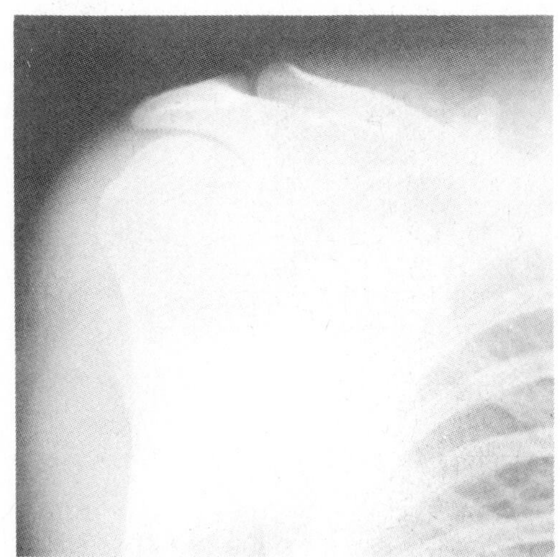

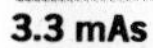

3.3 mAs

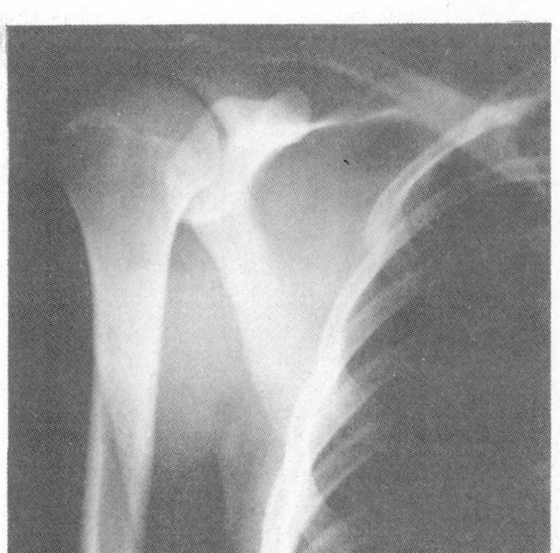

6.6 mAs

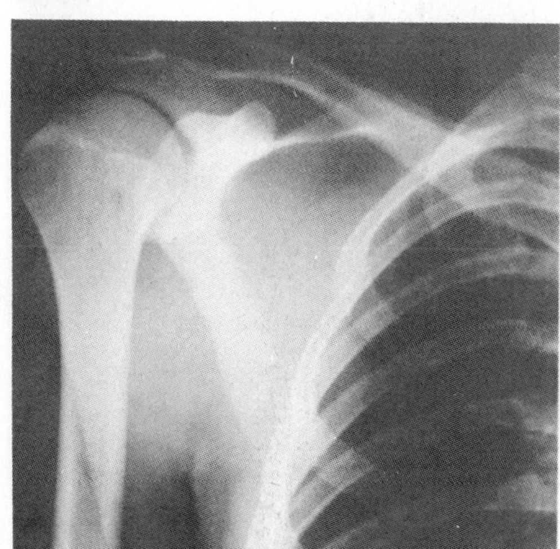

5 mAs

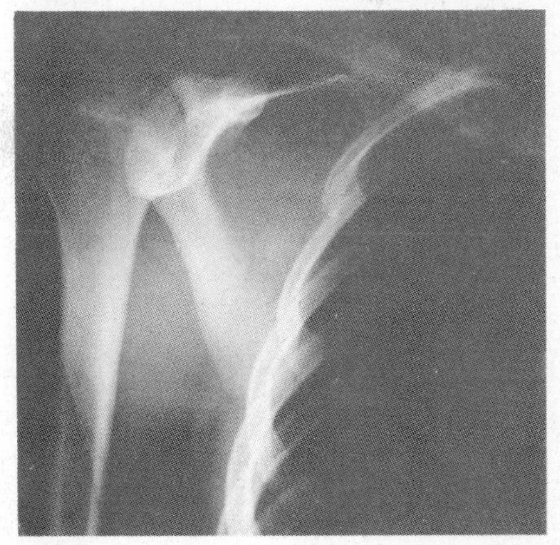

13.3 mAs

MILLIAMPERE TIME COMPUTATIONS

Impulses	Time	25 mA	50 mA	100 mA	150 mA	200 mA	300 mA	400 mA	500 mA	600 mA	700 mA	800 mA	1000 mA	1200 mA
1	1/120	.2	.4	.8	1.3	1.6	2.6	3.3	4.2	5	5.8	6.6	8	10
2	1/60	.4	.8	1.6	2.5	3.3	5	6.6	8.3	10	11.6	13.3	16	20
3	1/40	.6	1.2	2.5	3.7	5	7.5	10	12.5	15	17.5	20	25	30
4	1/30	.8	1.6	3.3	5	6.6	10	13.3	16.6	20	23.3	26.6	33	40
5	1/24	1	2.1	4.2	6.2	8.3	12.5	16.6	21	25	29	33.3	41	50
6	1/20	1.2	2.5	5	7.5	10	15	20	25	30	35	40	50	60
8	1/15	1.6	3.3	6.6	10	13.3	20	26.6	33.3	40	46.6	53.3	66	80
10	1/12	2.1	4.2	8.3	12.5	16.6	25	33.3	41.6	50	58.3	66.6	83	100
12	1/10	2.5	5	10	15	20	30	40	50	60	70	80	100	120
15	1/8	3.1	6.2	12.5	18.7	25	37.5	50	62.5	75	87.5	100	125	150
16	2/15	3.3	6.6	13.3	20	26.6	40	53.3	66.6	80	93.3	106.6	133	160
18	3/20	3.7	7.5	15	22.5	30	45	60	75	90	105	120	150	180
20	1/6	4.2	8.3	16.6	25	33.3	50	66.7	83.3	100	116.6	133	166	200
24	1/5	5	10	20	30	40	60	80	100	120	140	160	200	240
30	1/4	6.2	12.5	25	37.5	50	75	100	125					
32	4/15	6.6	13.3	26.6	40	53.3	80	106.6	133.3					
36	3/10	7.5	15	30	45	60	90	120	150					
	4/10	10	20	40	60	80	120							
	5/10	12.5	25	50	75	100	150							
	6/10	15	30	60	90	120	180							
	7/10	17.5	35	70	105	140	210							
	3/4	18.7	37.5	75	112.5	150	225							
	8/10	20	40	80	120	160	240							
	9/10	22.5	45	90	135	180	270							
	1	25	50	100	150	200	300							
	1- 1/4	31.2	62.5	125	187.5	250	375							
	1- 1/2	37.5	75	150	225	300	450							
	1- 3/4	43.7	87.5	175	262.5	350	525							
	2	50	100	200	300	400	600							
	2- 1/4	56.2	112.5	225										
	2- 1/2	62.5	125	250										
	2- 3/4	68.7	137.5	275										
	3	75	ı50	300										
	3- 1/4	81.2	162.5	325										
	3- 1/2	87.5	175	350										
	3- 3/4	93.7	187.5	375										
	4	100	200	400										

Time	100 mA	200 mA	300 mA	400 mA	500 mA	600 mA	700 mA	800 mA	1000 mA	1200 mA	1400 mA	1500 mA	1600 mA	1800 mA	2000 mA
.001	.1	.2	.3	.4	.5	.6	.7	.8	1	1.2	1.4	1.5	1.6	1.8	2
.002	.2	.4	.6	.8	1	1.2	1.4	1.6	2	2.4	2.8	3	3.2	3.6	4
.003	.3	.6	.9	1.2	1.5	1.8	2.1	2.4	3	3.6	4.2	4.5	4.8	5.4	6
.004	.4	.8	1.2	1.6	2.0	2.4	2.8	3.2	4	4.8	5.6	6	6.4	7.2	8
.005	.5	1	1.5	2.0	2.5	3.0	3.5	4.0	5	6	7	7.5	8	9	10
.006	.6	1.2	1.8	2.4	3.0	3.6	4.2	4.8	6	7.2	8.4	9	9.6	10.2	12
.008	.8	1.6	2.4	3.2	4.0	4.8	5.6	6.4	8	9.6	11.2	12	12.8	14.2	16
.01	1	2	3	4	5	6	7	8	10	12	14	15	16	18	20
.02	2	4	6	8	10	12	14	16	20	24	28	30	32	36	40
.03	3	6	9	12	15	18	21	24	30	36	42	45	48	54	60
.04	4	8	12	16	20	24	28	32	40	48	56	60	64	72	80
.05	5	10	15	20	25	30	35	40	50	60	70	75	80	90	100
.06	6	12	18	24	30	36	42	48	60	72	84	90	96	102	120
.08	8	16	24	32	40	48	56	64	80	96	112	120	128	142	160
.10	10	20	30	40	50	60	70	80	100	120	140	150	160	180	200
.12	12	24	36	48	60	72	84	96	120	144	168	180	192	216	240
.16	16	32	48	64	80	96	112	178	160	192	224	240	256	288	320
.20	20	40	60	80	100	120	140	160	200	240	280	300	320	360	400
.25	25	50	75	100	125	150	175	200	250	300	350	365	400	450	500
.32	32	64	96	128	160	192	224	256	320	384	448	480	512		
.40	40	80	120	160	200	240	280	320	400	480					
.50	50	100	150	200	250	300	350	400	500						
.64	64	128	192	296	320	384	448	512							
.80	80	160	240	320	480	480									
1	100	200	300	400	500										
1.2	120	240	360	480											
1.5	150	300													
2	200	400													
2.5	250	500													
3.2	320														
4	400														
5	500														

SUGGESTED TECHNIQUES

FINGERS

One Detail Screen
Table Top
40 inch (102 cm) Focal-Film Distance

kVp 60

cm. Range	mAs
1-3	5

Measuring Reference:
Measure the thickness through the base of the phalanges.

HAND

One Detail Screen
Table Top
40 inch (102 cm) Focal-Film Distance

kVp 60

Anterior—Oblique

cm. Range	mAs
1-2	6.6
3-5	10
6-8	12.5

Lateral

cm. Range	mAs
4-6	15
7-10	20
11-13	25

Measuring References:
Anterior—Measure the thickness through the head of the third metacarpal.

Oblique—Use the same technique that was used for the anterior view.

Lateral—Measure the thickness through the head of the second metacarpal.

WRIST

One Detail Screen
Table Top
40 inch (102 cm) Focal-Film Distance

kVp 60

Anterior—Oblique

cm. Range	mAs
1-2	6.6
3-6	10
7-10	12.5

Lateral

cm. Range	mAs
3-4	15
5-8	20
9-11	25

Measuring References:
Anterior—Measure the thickness through the midpoint of the styloid processes.

Oblique—Use the same technique that was used for the anterior view.

Lateral—Measure the thickness through the lateral styloid process.

FOREARM

One Detail Screen
Table Top
40 inch (102 cm) Focal-Film Distance

kVp 60

cm. Range	mAs
3-5	10
6-9	12.5
10-12	15

Measuring Reference:

Measure the thickness of the forearm at the point traversed by the Central Ray on all views.

ELBOW

One Detail Screen
Table Top
40 inch (102 cm) Focal-Film Distance

kVp 60

Posterior

cm. Range	mAs
3-5	15
6-9	20
10-12	25

Lateral

cm. Range	mAs
4-6	15
7-10	20
11-13	25

Measuring References:
Posterior—Measure the thickness through the midpoint of the epicondyles.

Lateral—Measure the thickness through the lateral epicondyle.

TOES

One Detail Screen
Table Top
40 inch (102 cm) Focal-Film Distance

kVp 60

cm. Range	mAs
1-2	5
3-5	6.6

Measuring Reference:
Measure the thickness through the base of the phalanges.

FOOT

One Detail Screen
Table Top
40 inch (102 cm) Focal-Film Distance

kVp 60

Posterior—Oblique

cm. Range	mAs
3-5	10
6-9	12.5
10-13	15

Lateral

cm. Range	mAs
3-5	15
6-9	20
10-13	25

Measuring References:
Posterior—Measure the thickness through the superior bony prominence of the foot.

Oblique—Use the same technique that was used for the posterior view.

Lateral—Measure the thickness through the head of the first metatarsal.

OS CALCIS

One Detail Screen
Table Top
40 inch (102 cm) Focal-Film Distance

kVp 60

Axial

cm. Range	mAs
4-6	25
7-10	30
11-14	40

Lateral

cm. Range	mAs
3-5	15
6-9	20
10-12	25

Measuring References:
Axial—The Os Calcis is difficult to measure. Accordingly, measure the patient as you would for a posterior view of the ankle.

Lateral—Measure the thickness through a point midway between the medial malleolus and the plantar surface of the foot.

ANKLE

One Detail Screen
Table Top
40 inch (102 cm) Focal-Film Distance

kVp 60

Posterior—Oblique

cm. Range	mAs
4-6	25
7-10	30
11-14	40

Lateral

cm. Range	mAs
3-5	15
6-9	20
10-13	25

Measuring References:
Posterior—Measure the thickness through the midpoint of the malleoli.

Oblique—Use the same technique that was used for the posterior view.

Lateral—Measure the thickness through the medial malleolus.

HUMERUS

Intensifying Screens
12:1 Grid Ratio
40 inch (102 cm) Focal-Film Distance

kVp 80

Posterior in External and Internal Rotation

cm. Range	Hi-Plus mAs	Par Speed mAs
8-12	5	10
13-16	6.6	13.3
17-20	10	20

kVp 80

Trans-Thoracic View

cm. Range	Hi-Plus mAs	Par Speed mAs
28-29	60	125
30-31	80	150
32-33	100	200
34-35	125	250
36-37	150	300
38-40	200	400

Measuring References:
Posterior—Measure the thickness of the arm at the point traversed by the Central Ray on all views.

Trans-Thoracic—Measure the thickness from the upper third of the arm through the thorax.

SHOULDER

Intensifying Screens
12:1 Grid Ratio
40 inch (102 cm) Focal-Film Distance

kVp 80

Posterior in External and Internal Rotation

cm. Range	Hi-Plus mAs	Par Speed mAs
8-11	5	10
12-16	6.6	13.3
17-20	10	20

Measuring Reference:
Posterior—Measure the thickness through the coracoid process.

CLAVICLE

Intensifying Screens
12:1 Grid Ratio
40 inch (102 cm) Focal-Film Distance

kVp 80

Anterior—Axial View

cm. Range	Hi-Plus mAs	Par Speed mAs
9-12	5	10
13-17	6.6	13.3
18-21	10	20

Measuring References:

Anterior—Measure the thickness at the midpoint of the body of the clavicle.

Axial—Use the same technique that was used for the anterior view.

SCAPULA

Intensifying Screens
12:1 Grid Ratio
40 inch (102 cm) Focal-Film Distance

kVp 80

Posterior

cm. Range	Hi-Plus mAs	Par Speed mAs
8-12	5	10
12-16	6.6	13.3
18-22	10	20

Lateral

cm. Range	Hi-Plus mAs	Par Speed mAs
14-17	6.6	13.3
18-22	10	20
23-27	16.6	33.3

Measuring References:

Posterior—Measure the thickness through the coracoid process.

Lateral—With the patient in position, measure the thickness through the midpoint of the axillary border. The measurement is obtained by measuring the thickness between the axillary and vertebral borders of the scapula.

STERNUM

Intensifying Screens
12:1 Grid Ratio
40 inch (102 cm) Focal-Film Distance

kVp 80

Right Anterior Oblique

cm. Range	Hi-Plus mAs	Par Speed mAs
18-19	10	20
20-21	15	30
22-23	20	40
24-25	30	60
26-27	50	100
28-29	80	150
30-31	100	200

Lateral

cm. Range	Hi-Plus mAs	Par Speed mAs
24-25	15	30
26-27	25	50
28-29	30	60
30-31	40	80
32-33	50	100
34-35	60	120
36-37	80	150
38-39	100	200

Measuring References:

Right Anterior Oblique—With the patient in position, measure the thickness through the midpoint of the body of the sternum.

Lateral—With the patient in position, measure the thickness through the midpoint of the body of the sternum. Note: The entire thorax is measured.

RIBS

Intensifying Screens
12:1 Grid Ratio
40 inch (102 cm) Focal-Film Distance

Ribs Above the Diaphragm

kVp 80

Posterior—Anterior—Oblique

cm. Range	Hi-Plus mAs	Par Speed mAs
14-15	4.2	8.3
16-17	6.6	13.3
18-19	8.3	16.6
20-21	12.5	25
22-23	15	30
24-25	20	40
26-27	30	60
28-29	40	80
30-31	50	100

Ribs Below the Diaphragm

kVp 80

Posterior—Anterior—Oblique

cm. Range	Hi-Plus mAs	Par Speed mAs
14-15	15	30
16-17	20	40
18-19	30	60
20-21	40	80
22-23	50	100
24-25	70	150
26-27	100	200
28-29	125	250
30-31	175	300

Measuring References:
Ribs Above the Diaphragm—Measure the thickness through the sixth thoracic vertebra.

Ribs Below the Diaphragm—Measure the thickness through the umbilicus.

120 KILOVOLTS PEAK CHEST TECHNIQUE

A high voltage grid technique for chest radiography has much to recommend it.* The latitude makes serial films consistently alike; duplication of films, for comparison, is relatively easy and the initial studies seldom need to be repeated because of poor technique.

The extended scale of radiographic contrast may be disconcerting at first, but the advantages over conventional lower voltage techniques are many:

a. One can see "behind" the heart on the anterior projection
b. The areas along the lateral chest walls, which are frequently obscure due to muscle tissue, are clearly seen.
c. The trachea and main bronchi are very clearly seen.
d. The bone detail of the thorax is seen more clearly than on low voltage chest films.
e. The vascular pattern of the lungs is well demonstrated. In some cases it is shown better than on comparable low voltage films.
f. The scale of radiographic contrast remains constant because the kilovoltage does not change per centimeter thickness of the patient.

The technique makes use of:

72 inch (183 cm) Focal-Film Distance
120 Kilovolts Peak
Variable Milliampere-Second Values
A Stationary Wafer Grid: 103 line, 6:1 ratio
28-72 inch (70-183 cm) focus

The use of automatic timers, such as the photoelectric timer (phototimers) and ionization timer is widely prevalent in chest radiography and will take into account patient variations. However, without automatic timers, we feel that variations in chest technique must be made according to variations in the physical stature of the patient. The technique described here is based on the variation of the milliampere-second values according to the build of the patient. The increase or decrease in milliampere seconds is controlled by the use of "type" factors as described on the next page. Changing techniques by 1 type is equal to a 1 centimeter change.

*This technique is based on the research of Charles D. Smith, M.D. and Finese H. Wilbourne, R.T., Department of Radiology, Roanoke Memorial Hospitals, Roanoke, Virginia.

Each patient is measured. The lateral measurement is made with the patient's arms extended laterally, the measuring bars of a caliper being placed as high as possible against the axillary folds. The anterior measurement is then made at the same level. This is important. Measurements must be made at the same level on every patient at this point.

Figure 1
The patient is in the correct position to be measured for the lateral view of the chest. Note that the forearms are parallel with the floor, and that the knuckles are touching.

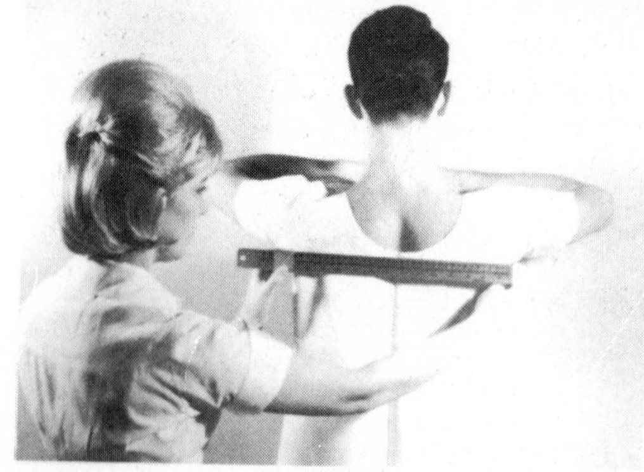

Figure 2
The technologist measures the patient on a parallel plane extending from the right to the left axilla for the lateral view of the chest.

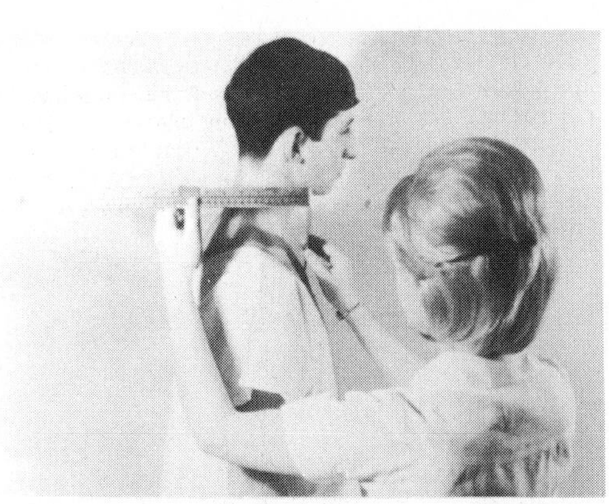

Figure 3
Note how the technologist is measuring the patient for the anterior view. The calipers are so placed that they measure through the thorax on the same plane as was used to obtain the lateral measurement.

To those uninitiated in chest radiography, the following "type" variations may seem needlessly complicated, but there is no easy method that will assure good chest films. People with short chests and little lung tissue require more milliampere-seconds than people with long chests and relatively large amounts of lung tissue, even though they may both have equal anterior centimeter measurements. The following "types" may be used:

CALCULATION OF THE ANTERIOR TECHNIQUE:

1. Use the anterior centimeter measurements to establish a base milliampere-seconds.

Breast Factor:

2. If the patient has large or relatively large breasts + 1 type.

Height and Weight Factor:

3. The next type depends on the height and weight of the patient and the weight is a factor if the patient is less than 5′4″ tall.
   a. If 5′4″ tall or less, and: 125-150 pounds + 1 type; over 150 pounds +2 types; over 200 pounds +3 types.
   b. If over 5′4″ tall and 200 pounds or over + 1 type.

Measurement Factor:

4. The next category is calculated by comparing the anterior and lateral measurements. A chest that is relatively wide compared with the anterior diameter is more muscular and will require higher milliampere-second values.

   If the lateral measurement is more than the anterior measurement by 10-14 centimeters +1 type; 15-19 centimeters +2 types; 20 (or more) centimeters +3 types.

Minus Types:

5. Patients with emaciation or emphysema will require lower milliampere-second values than the base values. When present −1 type, emaciation is relatively easy to diagnose, merely by observation.

CALCULATION OF THE LATERAL TECHNIQUE:

1. Measure the patient and use the technique indicated, except where minus types are indicated.

An example of patient typing may help to illustrate how this technique is implemented.

	Hi-Plus mAs
Anterior Chest:	
Our patient is a young woman measuring 19 cms	1.6
Breast Factor:	
Plus 1 type for breasts	2.5
Height and Weight Factor:	
It is not necessary to add types under this factor because the patient is 5′ 6″ tall and weighs 124 pounds.	
Measurement Factor:	
The lateral measurement is 31 centimeters which would indicate plusing 1 additional type	3.3
Minus Types:	
Emaciation or emphysema is not a factor with this patient, therefore, 3.3 milliampere-seconds would be used for the anterior view.	
Lateral Chest:	
As there are no minus types present, 31 centimeters would indicate using 8.3 milliampere-seconds for the lateral view.	

SUMMARY:

A fixed kilovoltage chest technique utilizing a 6:1 ratio grid has been described. The control of radiographic density is based upon the technologist carefully measuring each patient, and then typing the patient according to the body habitus. It has been our experience that this chest technique will give satisfactory results, and once mastered, is easily used by both graduate and student technologists.

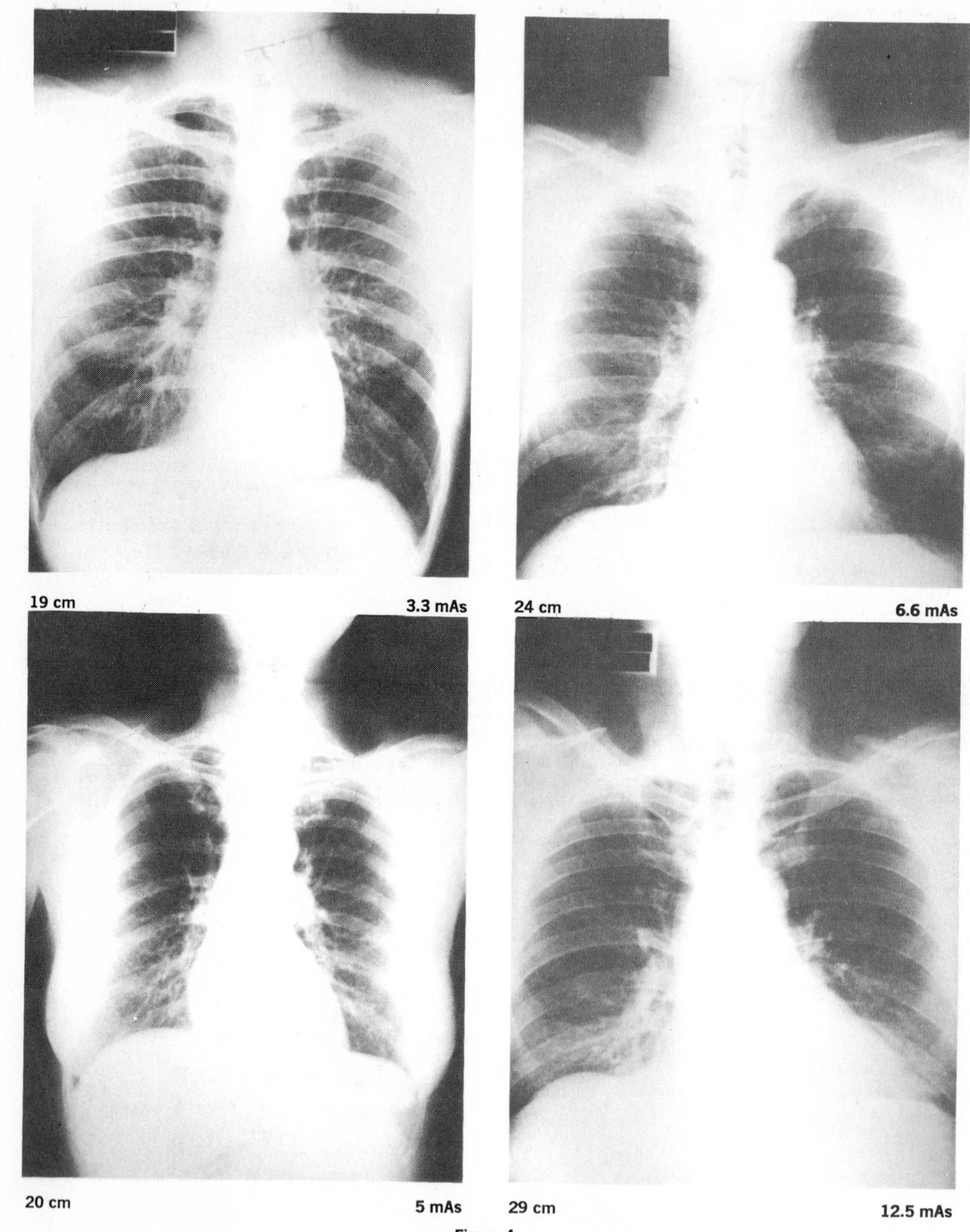

Figure 4
A series of anterior chest radiographs on different patients utilizing the 120 kVp technique.

120 KILOVOLTS PEAK CHEST TECHNIQUE

Range: 6 years through to adults. Children and adults who do not measure anterior 19 centimeters and lateral 24 centimeters start typing from the Base Line Factor (shown in **bold type**).

ANTERIOR CHEST

cm. Range	Hi-Plus mAs	Par Speed mAs
19	**1.6**	**2.5**
20	2.5	4.2
21	3.3	5
22	4.2	7.5
23	5	8.3
24	5	8.3
25	6.6	10
26	6.6	10
27	7.5	12.5
28	8.3	13.3
29	10	16.6
30	12.5	20

LATERAL CHEST

cm. Range	Hi-Plus mAs	Par Speed mAs
24	**4.2**	**7.5**
25	5	8.3
26	5	8.3
27	6.6	10
28	6.6	10
29	7.5	12.5
30	8.3	13.3
31	8.3	13.3
32	10	16.6
33	12.5	20
34	13.3	25
35	16.6	30
36	20	40
37	25	50
38	30	50
39	33.3	60
40	40	70

OBLIQUE CHEST

+2 types over the anterior technique

LORDOTIC CHEST

+4 types over the anterior technique

CHEST

Recumbent with Potter-Bucky Diaphragm

Intensifying Screens
12:1 Grid Ratio
40 inch (102 cm) Focal-Film Distance

When the patient cannot stand or sit, and a routine chest technique is required:

kVp 100

a. Use the 120 kilovolts peak, 72 inch (183 cm) focal-film distance chest technique milliampere-second values, pg. 13.
b. Measure the patient, and use the milliampere-second value indicated.

OVERPENETRATED VIEW OF CHEST:

kVp 100

a. Use the 120 kilovolts peak, 72 inch (183 cm) focal-film distance chest technique milliampere-second values, pg. 13.
b. Measure the patient, and +4 types.

BRONCHOGRAM:

kVp 100

a. Use the 120 kilovolts peak, 72 inch (183 cm) focal-film distance chest technique milliampere-second values, pg. 13.
b. Measure the patient, and +3 types.

Measuring Reference:
On all views, measure the thickness through the sixth thoracic vertebra.

NON-GRID CHEST TECHNIQUE

Intensifying Screens
72 inch (183 cm) Focal-Film Distance

kVp 80
Anterior

cm. Range	Hi-Plus mAs	Par Speed mAs
16-17	1.6	3.3
18-19	2.5	5
20-21	3.3	6.6
22-23	4.2	8.3
24-25	5	10
26-27	6.6	13.3
28-29	8.3	16.6

kVp 80
Oblique

cm. Range	Hi-Plus mAs	Par Speed mAs
22-23	6.6	13.3
24-25	8.3	16.6
26-27	10	20
28-29	13.3	25
30-31	16.6	30
32-33	20	40
34-35	25	50

kVp 90
Lateral

cm. Range	Hi-Plus mAs	Par Speed mAs
22-23	4.2	8.3
24-25	5	10
26-27	6.6	13.3
28-29	8.3	16.6
30-31	10	20
32-33	13.3	25
34-35	16.6	30
36-37	20	40
38-39	25	50
40-41	30	60

kVp 80
Lordotic

cm. Range	Hi-Plus mAs	Par Speed mAs
18-19	4.2	8.3
20-21	5	10
22-23	6.6	13.3
24-25	8.3	16.6
26-27	10	20

Measuring Reference:
With the patient in position, measure the thickness through the thorax at the level of the sixth thoracic vertebra.

AIR GAP CHEST TECHNIQUE

Intensifying Screens
120 inch (305 cm) Focal-Film Distance

Non-Grid
6 inch (15 cm) Air Gap

kVp 120

Anterior

cm. Range	Hi-Plus mAs	Par Speed mAs
16-17	1.6	3.3
18-19	2.5	5
20-21	3.3	6.6
22-23	4.2	8.3
24-25	5	10
26-27	6.6	13.3
28-29	8.3	16.6

Oblique

cm. Range	Hi-Plus mAs	Par Speed mAs
22-23	5	10
24-25	6.6	13.3
26-27	8.3	16.6
28-29	10	20
30-31	13.3	25
32-33	16.6	30
34-35	20	40

Lateral

cm. Range	Hi-Plus mAs	Par Speed mAs
22-23	3.3	6.6
24-25	4.2	8.3
26-27	5	10
28-29	6.6	13.3
30-31	8.3	16.6
32-33	10	20
34-35	13.3	25
36-37	16.6	30
38-39	20	40
40-41	25	50

Lordotic

cm. Range	Hi-Plus mAs	Par Speed mAs
18-19	4.2	8.3
20-21	5	10
22-23	6.6	13.3
24-25	8.3	16.6
26-27	10	20

Measuring Reference:

With the patient in position, measure the thickness through the thorax at the level of the sixth thoracic vertebra.

ABDOMEN

(UROGRAM—GALL BLADDER)

Intensifying Screens
12:1 Grid Ratio
40 inch (102 cm) Focal-Film Distance

kVp 80

Posterior—Oblique—Lateral—Decubitus

cm. Range	Hi-Plus mAs	Par Speed mAs
12-13	10	20
14-15	15	30
16-17	20	40
18-19	30	60
20-21	40	80
22-23	50	100
24-25	70	150
26-27	100	200
28-29	125	250
30-31	175	300
32-33	225	400

POSTBARIUM STUDIES: +10 kVp

Measuring References:

Posterior—Anterior—Oblique—Decubitus—Lateral—Measure the thickness through the umbilicus.

Upright—Measure the thickness through the apex of the abdomen in the upright position.

PELVIMETRY

Intensifying Screens
12:1 Grid Ratio
36 inch (91 cm) Focal-Film Distance

kVp 120

Posterior—Lateral

cm. Range	Hi-Plus mAs	Par Speed mAs
18-19	4.2	8.3
20-21	6.6	13.3
22-23	8.3	16.6
24-25	10	20
26-27	15	30
28-29	20	40
30-31	30	60
32-33	40	80
34-35	50	100
36-37	60	120
38-39	80	150
40-41	100	200

NOTE: This technique is based on the use of the Snow Obstetrical Calculator for measurements.

Measuring References:

Posterior—Measure the thickness through the apex of the abdomen.

Lateral—Measure the thickness through a point directly superior to the crests of the ilium.

BARIUM STUDIES

Intensifying Screens
12:1 Grid Ratio
40 inch (102 cm) Focal-Film Distance

	kVp
Gastrointestinal Series	120
Barium Enema	120
Air Contrast Studies	100

All Views

cm. Range	Hi-Plus mAs	Par Speed mAs
16-17	8.3	15
18-20	10	20
21-22	15	30
23-24	20	40
25-27	30	60
28-31	40	80
32-35	60	100
36-38	80	150

Measuring Reference:

Measure the thickness of the anatomy at the point traversed by the Central Ray on all views.

LEG

One Detail Screen
Table Top
40 inch (102 cm) Focal-Film Distance

kVp 60

Posterior

cm. Range	mAs
6-8	40
9-12	60
13-15	80

Lateral

cm. Range	mAs
5-7	25
8-11	30
12-15	40

Measuring Reference:

Measure the thickness of the leg at the point traversed by the Central Ray on all views.

KNEE

Intensifying Screens
12:1 Grid Ratio
40 inch (102 cm) Focal-Film Distance

kVp 80

Anterior—Posterior

cm. Range	Hi-Plus mAs	Par Speed mAs
7-9	3.3	6.6
10-14	5	10
15-17	6.6	13.3

Lateral

cm. Range	Hi-Plus mAs	Par-Speed mAs
6-8	2.5	5
9-13	3.3	6.6
14-15	5	10

Homblad View (Tunnel View)

cm. Range	Hi-Plus mAs	Par-Speed mAs
7-9	5	10
10-14	7.5	15
15-17	10	20

NOTE: If it is desired to do the knee routine with screens table top:

1—Reduce the kilovoltage to 60.
2—Use the milliampere-second values indicated above.

Measuring References:

Anterior—Posterior—Homblad—Measure the thickness through the apex of the patella.

Lateral—Measure the thickness through the coronal plane at the apex of the patella.

PATELLA

One Detail Screen
Table Top
40 inch (102 cm) Focal-Film Distance
Extension Cone

kVp 60

Anterior—Axial

cm. Range	mAs
7-9	40
10-14	60
15-17	80

Lateral

cm. Range	mAs
6-8	30
9-13	40
14-16	60

Measuring References:

Anterior—Measure the thickness through the apex of the patella.

Axial View—Use the same technique that was used for the anterior view.

Lateral—Measure the thickness through the coronal plane at the apex of the patella.

FEMUR

Intensifying Screens
12:1 Grid Ratio
40 inch (102 cm) Focal-Film Distance

kVp 80

Distal: Posterior—Lateral

cm. Range	Hi-Plus mAs	Par Speed mAs
9-12	7.5	15
13-15	10	20
16-20	15	30

Proximal: Posterior—Lateral

cm. Range	Hi-Plus mAs	Par Speed mAs
10-14	10	20
15-17	15	30
18-22	20	40

Measuring Reference:

Measure the thickness of the thigh at the point traversed by the Central Ray on all views.

HIP

Intensifying Screens
40 inch (102 cm) Focal-Film Distance
Extension Cone

kVp 80

Posterior
12:1 Grid Ratio

cm. Range	Hi-Plus mAs	Par-Speed mAs
8-9	5	10
10-11	7.5	13.3
12-13	8.3	16.6
14-15	10	20
16-17	15	30
18-19	20	40
20-21	25	50
22-23	30	60

Oblique Lateral (Frog Position)
12:1 Grid Ratio

cm. Range	Hi-Plus mAs	Par Speed mAs
12-13	10	20
14-15	15	30
16-17	20	40
18-19	25	50
20-21	30	60
22-23	40	80
24-25	50	100
26-27	60	150

True Lateral
8:1 Grid Ratio

cm. Range	Hi-Plus mAs	Par Speed mAs
14-15	40	80
16-19	50	100
20-21	60	125
22-25	80	150
26-29	100	200

NOTE: The true lateral projection is taken with the x-ray tube directed across the table; an 8:1 wafer grid is taped to a cassette.

Measuring References:

Posterior—Oblique Lateral—Measure the thickness through the intersecting lines of the anterosuperior iliac spine and the symphysis pubis.

True Lateral—Measure the thickness from the medial aspect of the thigh through to the greater-trochanter.

CERVICAL SPINE

Intensifying Screens

kVp 80

12:1 Grid Ratio

40 inch (102 cm) Focal-Film Distance

Posterior—Oblique

cm. Range	Hi-Plus mAs	Par Speed mAs
8-9	4.2	7.5
10-11	5	10
12-13	7.5	15
14-15	10	20
16-17	15	30

Open Mouth View of C1 and C2—Collimate to a 4 × 4 inch (10 × 10 cm) field

cm. Range	Hi-Plus mAs	Par Speed mAs
8-9	5	10
10-11	7.5	15
12-13	10	20
14-15	15	30
16-17	20	40

kVp 80

6:1 Grid Ratio

72 inch (183 cm) Focal-Film Distance

Oblique—Lateral

cm. Range	Hi-Plus mAs	Par Speed mAs
6-7	7.5	15
8-9	10	20
10-11	13.3	25
12-13	16.6	30
14-15	20	40
16-17	30	60

Measuring Reference:

Measure the thickness at the level of the thyroid cartilage on all views.

CERVICO-THORACIC SPINE

Intensifying Screens
12:1 Grid Ratio
40 inch (102 cm) Focal-Film Distance

kVp 80

Fletcher View

cm. Range	Hi-Plus mAs	Par Speed mAs
20-24	50	100
25-28	100	200
29-33	200	400

Measuring Reference:

Measure the thickness through the seventh cervical vertebrá in the same plane as traversed by the Central Ray.

THORACIC SPINE

Intensifying Screens
12:1 Grid Ratio
40 inch (102 cm) Focal-Film Distance

kVp 80

Posterior

cm. Range	Hi-Plus mAs	Par Speed mAs
14-15	15	30
16-17	20	40
18-19	25	50
20-21	30	60
22-23	40	80
24-25	50	100
26-27	70	150
28-29	100	200
30-31	125	250

Lateral

cm. Range	Hi-Plus mAs	Par Speed mAs
20-21	10	20
22-23	15	30
24-25	20	40
26-27	30	60
28-29	40	80
30-31	50	100
32-33	60	120
34-35	80	150
36-37	100	200
38-39	150	300

Measuring Reference:

Measure the thickness through the sixth thoracic vertebra on all views.

LUMBAR SPINE

Intensifying Screens
12:1 Grid Ratio
40 inch (102 cm) Focal-Film Distance

kVp 80

Posterior

cm. Range	Hi-Plus mAs	Par Speed mAs
12-13	10	20
14-15	15	30
16-17	20	40
18-19	30	60
20-21	40	80
22-23	50	100
24-25	80	150
26-27	100	200
28-29	150	300
30-31	200	400

Oblique

cm. Range	Hi-Plus mAs	Par Speed mAs
12-13	15	30
14-15	20	40
16-17	25	50
18-19	30	60
20-21	40	80
22-23	60	100
24-25	100	200
26-27	125	250
28-29	150	300

kVp 100

Lateral

cm. Range	Hi-Plus mAs	Par Speed mAs
22-23	20	40
24-25	30	60
26-27	40	80
28-29	60	120
30-31	80	150
32-33	100	200
34-35	125	250
36-37	150	300
38-39	200	400
40-41	225	450

Measuring References:

Posterior—Measure the thickness through the umbilicus.

Oblique—With the patient in position, measure the thickness through the umbilicus.

Lateral—Measure the thickness through a point directly superior to the crests of the ilium.

LUMBAR SPINE Spot of L5

Intensifying Screens
12:1 Grid Ratio
40 inch (102 cm) Focal-Film Distance
Extension Cone

kVp 100

cm. Range	Hi-Plus mAs	Par Speed mAs
26-27	60	120
28-29	80	150
30-31	100	200
32-33	125	250
34-35	150	300
36-37	200	400
38-39	250	500
40-41	300	600

Measuring Reference:

Spot of L5—Measure the thickness through the fifth lumbar vertebra.

PELVIS

Intensifying Screens
12:1 Grid Ratio
40 inch (102 cm) Focal-Film Distance

kVp 80

Posterior

cm. Range	Hi-Plus mAs	Par Speed mAs
12-13	10	20
14-15	15	30
16-17	20	40
18-19	30	60
20-21	40	80
22-23	50	100
24-25	80	150
26-27	100	200

kVp 100

Lateral

cm. Range	Hi-Plus mAs	Par Speed mAs
22-23	20	40
24-25	30	60
26-27	40	80
28-29	60	125
30-31	80	150
32-33	100	200
34-35	125	250
36-37	150	300
38-39	200	400
40-41	225	450

Measuring Reference:

Measure the thickness through the soft depression below the anterosuperior iliac spine on all views.

SACRO-ILIAC ARTICULATION

Intensifying Screens
12:1 Grid Ratio
40 inch (102 cm) Focal-Film Distance
Extension Cone

kVp 80

Oblique

cm. Range	Hi-Plus mAs	Par Speed mAs
12-13	15	30
14-15	20	40
16-17	25	50
18-19	30	60
20-21	40	80
22-23	60	120
24-25	100	200
26-27	125	250
28-29	150	300

Measuring Reference:

Oblique—With the patient in position, measure the thickness through a point 1 inch (2.5 cm) medial to the elevated superior iliac spine.

SACRUM-COCCYX

Intensifying Screens
12:1 Grid Ratio
40 inch (102 cm) Focal-Film Distance

kVp 80

Posterior

cm. Range	Hi-Plus mAs	Par Speed mAs
12-13	10	20
14-15	15	30
16-17	20	40
18-19	30	60
20-21	40	80
22-23	50	100
24-25	80	150
26-27	100	200

kVp 100

Lateral

cm. Range	Hi-Plus mAs	Par Speed mAs
20-21	15	30
22-23	20	40
24-25	30	60
26-27	40	80
28-29	50	100
30-31	70	150
32-33	100	200
34-35	125	250
36-37	150	300
38-39	200	400
40-41	250	500

Measuring References:

Posterior—Measure the thickness through a point midway between the superior iliac spine and the symphysis pubis.

Sacrum Lateral—Measure the thickness at the level of the anterosuperior iliac spine.

Coccyx Lateral—Measure the thickness in the same plane as traversed by the Central Ray.

SKULL

Intensifying Screens
8:1 Grid Ratio
36 inch (91 cm) Focal-Film Distance

kVp 80

Chamberlin—Towne View

cm. Range	Hi-Plus mAs	Par Speed mAs
13-17	15	30
18-21	20	40
22-26	30	60

Anterior—Posterior—Waters View

cm. Range	Hi-Plus mAs	Par Speed mAs
13-17	10	20
18-21	15	30
22-26	20	40

Lateral

cm. Range	Hi-Plus mAs	Par Speed mAs
8-12	5	10
13-17	7.5	15
18-22	10	20

Basal View

cm. Range	Hi-Plus mAs	Par Speed mAs
15-19	30	60
20-25	50	100
26-29	60	120

Measuring References:

Chamberlin—Towne—Measure the thickness between the glabella and inion.

Anterior—Posterior—Measure the thickness between the glabella and inion.

Waters—Measure the thickness between the acanthion and the inion.

Lateral—Measure the thickness through a point directly superior to the ears.

Basal—Measure the thickness from the angle of the mandible to the vertex of the skull.

Note: These techniques are formulated for a head unit using a 36 inch (91 cm) focal-film distance and 8:1 ratio grid.

SINUSES

Intensifying Screens
8:1 Grid Ratio
36 inch (91 cm) Focal-Film Distance
Extension Cone

kVp 80

Caldwells View

cm. Range	Hi-Plus mAs	Par Speed mAs
13-17	15	30
18-21	20	40
22-26	30	60

Waters View

cm. Range	Hi-Plus mAs	Par Speed mAs
13-17	15	30
18-22	20	40
23-26	30	60

Lateral

cm. Range	Hi-Plus mAs	Par Speed mAs
9-12	2.5	5
13-17	3.3	6.6
18-21	5	10

Basal

cm. Range	Hi-Plus mAs	Par Speed mAs
15-19	30	60
20-25	50	100
26-29	60	120

Measuring References:

Caldwells—Measure the thickness between the glabella and inion.

Waters—Measure the thickness between the acanthion and the inion.

Lateral—Measure the thickness between the external canthi of the eye.

Basal—Measure the thickness from the angle of the mandible to the vertex of the skull.

MASTOIDS

Intensifying Screens
8:1 Grid Ratio
36 inch (91 cm) Focal-Film Distance
Extension Cone

kVp 80

Chamberlin-Towne View

cm. Range	Hi-Plus mAs	Par Speed mAs
13-17	20	40
18-21	30	60
22-26	40	80

Laws View

cm. Range	Hi-Plus mAs	Par Speed mAs
9-12	7.5	15
13-17	10	20
18-20	15	30

Stenvers View

cm. Range	Hi-Plus mAs	Par Speed mAs
13-14	5	10
15-19	10	20
20-24	15	30

Measuring References:

Chamberlin-Towne—Measure the thickness between the glabella and inion.

Laws—Measure the thickness from a point 2 inches (5 cm) superior to the ear on the up side to the mastoid tip under consideration.

Stenvers—Measure the thickness from the external canthus of the eye to the inion. To do this correctly, the median plane must be rotated forty-five degrees from the vertical.

SELLA TURCICA

Intensifying Screens
8:1 Grid Ratio
36 inch (91 cm) Focal-Film Distance
Extension Cone

kVp 80

Anterior

cm. Range	Hi-Plus mAs	Par Speed mAs
13-17	15	30
18-21	20	40
22-26	30	60

Lateral

cm. Range	Hi-Plus mAs	Par Speed mAs
9-12	5	10
13-17	7.5	15
18-21	10	20

Measuring References:

Anterior—Measure the thickness beween the glabella and inion.

Lateral—Measure the thickness through a point directly superior to the ears.

OPTIC FORAMEN

Intensifying Screens
8:1 Grid Ratio
36 inch (91 cm) Focal-Film Distance
Extension Cone

kVp 80

Rheese View

cm. Range	Hi-Plus mAs	Par Speed mAs
13-17	15	30
18-21	20	40
22-26	30	60

Measuring Reference:

Measure the thickness between the glabella and inion.

TEMPORO-MANDIBULAR JOINT

Intensifying Screens
8:1 Grid Ratio
36 inch (91 cm) Focal-Film Distance
Extension Cone

kVp 80

Semiaxial Transcranial View

cm. Range	Hi-Plus mAs	Par Speed mAs
9-12	7.5	15
13-17	10	20
18-20	15	30

Measuring Reference:

Measure the thickness from a point 2 inches (5 cm) superior to the ear on the up side to the temporo-mandibular joint under consideration.

FACIAL BONES

Intensifying Screens
8:1 Grid Ratio
36 inch (91 cm) Focal-Film Distance

kVp 80

Waters View

cm. Range	Hi-Plus mAs	Par Speed mAs
13-17	10	20
18-21	15	30
22-26	20	40

Lateral

cm. Range	Hi-Plus mAs	Par Speed mAs
9-12	2.5	5
13-17	3.3	6.6
18-21	5	10

Measuring References:

Waters—Measure the thickness between the acanthion and the inion.

Lateral—Measure the thickness between the external canthi of the eye.

ZYGOMATIC ARCH

Intensifying Screens
8:1 Grid Ratio
36 inch (91 cm) Focal-Film Distance
Extension Cone

kVp 80

Superior—Inferior View

cm. Range	Hi-Plus mAs	Par Speed mAs
9-11	5	10
12-17	7.5	15
18-21	10	20

Measuring Reference:

The zygomatic arch is difficult to measure. Accordingly, measure the patient as you would for a lateral view of the sinus.

NASAL BONES

One Detail Screen
Table Top
36 inch (91 cm) Focal-Film Distance
Extension Cone

kVp 60

Lateral

cm Range	mAs
-1	5
1.5-3	8.3

Measuring Reference:

Measure the thickness of the nasal bones at the point traversed by the Central Ray on all views.

MANDIBLE

Intensifying Screens
36 inch (91 cm) Focal-Film Distance

kVp 80
8:1 Grid Ratio
Anterior

cm. Range	Hi-Plus mAs	Par Speed mAs
10-12	10	20
13-18	15	30
19-22	20	40

kVp 60
Table Top
Extension Cone
Lateral

cm. Range	Hi-Plus mAs	Par Speed mAs
6-9	2.5	5
10-14	3.3	6.6
15-17	5	10

Measuring References:

Anterior—Measure the thickness between the mental point and the nape of the neck.

Lateral—Measure the thickness between the angles of the mandible.

A PEDIATRIC TECHNIQUE CHART

The July 1966 issue of "Radiologic Technology" contained an article on pediatric exposure factors by Clarence W. Heinlein, R.T. This technique gives excellent results, and based on his work, here is a suggested pediatric technique chart.

To convert these techniques to your department, the method of bracketed exposures can be used by making trial exposures at different kilovoltage levels. It is suggested that only one initial film be made to conserve exposure to ionizing radiation.

1. Measure or age type the child and make an exposure with the indicated kilovoltage.
2. If the trial film is overexposed, make a second exposure with a reduction of 4 kilovolts peak. If the trial film is underexposed, make a second exposure with an increase of 4 kilovolts peak.

One of the two trial films should give the desired results, and will indicate how much to change the kilovolt scales if necessary.

Milliampere and time combinations are suggested values only and can readily be changed to take advantage of higher capacity generators. It should be kept in mind though that a high speed Bucky is indicated for pediatric examinations to permit exposures down to 1/60 second.*

Like the adult technique chart, it has been found practical to present these techniques for use in a Kardex file. Measuring references when needed would be like those for adults.

*An alternative is to use a stationary 100 to 110 line grid in the Bucky.

PEDIATRIC TECHNIQUE

KILOVOLTAGE SCALES FOR VIEWS OF THE THORAX, SPINE, ABDOMEN, AND PELVIS

0 Months to 3 Months

Posterior													
cm.	5	6	7	8	9	10	11	12	13	14	15		
kVp	52	54	56	58	60	62	64	66	68	70	72		
Lateral													
cm.	10	11	12	13	14	15	16	17	18				
kVp	64	66	68	70	72	74	76	78	80				

3 Months to 2 Years

Posterior													
cm.	8	9	10	11	12	13	14	15	16	17	18		
kVp	62	64	66	68	70	72	74	76	78	80	82		
Lateral													
cm.	12	13	14	15	16	17	18	19	20	21	22		
kVp	72	74	76	78	80	82	84	86	88	90	92		

2 Years to 5 Years

Posterior													
cm.	10	11	12	13	14	15	16	17	18	19	20		
kVp	70	72	74	76	78	80	82	84	86	88	90		
Lateral													
cm.	14	15	16	17	18	19	20	21	22	23	24	25	26
kVp	80	82	84	86	88	90	92	94	96	98	100	102	104

5 Years to 12 Years

Posterior													
cm.	10	11	12	13	14	15	16	17	18	19	20		
kVp	74	76	78	80	82	84	86	88	90	92	94		
Lateral													
cm.	14	15	16	17	18	19	20	21	22	23	24	25	26
kVp	84	86	88	90	92	94	96	98	100	102	104	106	108

HAND

CRONEX® Detail Screens
Table Top
40 Inch (102 cm) Focal-Film Distance

	mAs	0 Mos. to 3 Mos. kVp	3 Mos. to 2 Years kVp	2 Yrs. to 5 Yrs. kVp	5 Yrs. to 12 Yrs. kVp
All	3.3	58	62	64	68

WRIST-FOREARM-ELBOW-FOOT

CRONEX Detail Screens
Table Top
40 Inch (102 cm) Focal-Film Distance

	mAs	0 Mos. to 3 Mos. kVp	3 Mos. to 2 Years kVp	2 Yrs. to 5 Yrs. kVp	5 Yrs. to 12 Yrs. kVp
All	3.3	62	66	68	72

SHOULDER-KNEE-FEMUR

CRONEX Hi-Plus Screens
12:1 Grid Ratio
40 Inch (102 cm) Focal-Film Distance

	mAs	0 Mos. to 3 Mos. kVp	3 Mos. to 2 Years kVp	2 Yrs. to 5 Yrs. kVp	5 Yrs. to 12 Yrs. kVp
All	6.6	62	66	70	74

HUMERUS-ANKLE-LEG

CRONEX Detail Screens
Table Top
40 Inch (102 cm) Focal-Film Distance

	mAs	0 Mos. to 3 Mos. kVp	3 Mos. to 2 Yrs. kVp	2 Yrs. to 5 Yrs. kVp	5 Yrs. to 12 Yrs. kVp
All	6.6	58	62	64	68

PELVIS-HIP

CRONEX Hi-Plus Screens
12:1 Grid Ratio
40 Inch (102 cm) Focal-Film Distance

USE PEDIATRIC KILOVOLTAGE SCALES (PAGE 24)

	mAs
Posterior	10
Frog View	10

Note: Measuring references would be like those for adults (see pages 17 and 19).

CERVICAL VERTEBRAE

CRONEX Hi-Plus Screens
12:1 Grid Ratio
40 Inch (102 cm) Focal-Film Distance

	mAs	0 Mos. to 3 Mos. kVp	3 Mos. to 2 Yrs. kVp	2 Yrs. to 5 Yrs. kVp	5 Yrs. to 12 Yrs. kVp
Posterior	6.6	66	70	74	78
Lateral	6.6	66	70	74	78
Lateral at 72″ (183 cm) FFD (Screen Exposure)	10	66	70	74	78

THORACIC-LUMBAR VERTEBRAE

CRONEX Hi-Plus Screens
12:1 Grid Ratio
40 Inch (102 cm) Focal-Film Distance

USE PEDIATRIC KILOVOLTAGE SCALES (PAGE 24)

	mAs
THORACIC	
Posterior	10
Lateral	10
LUMBOSACRAL	
Posterior	10
Lateral	20

Note: Measuring references would be like those for adults (see pages 18 and 19).

ABDOMEN

CRONEX Hi-Plus Screens
12:1 Grid Ratio
40 Inch (102 cm) Focal-Film Distance

USE PEDIATRIC KILOVOLTAGE SCALES (PAGE 24)

	mAs
All	10

Note: Measuring references would be like those for adults (see page 15).

CHEST

CRONEX Hi-Plus Screens
72 Inch (183 cm) Focal-Film Distance

	mAs	0 Mos. to 3 Mos. kVp	3 Mos. to 2 Yrs. kVp	2 Yrs. to 5 Yrs. kVp	5 Yrs. to 12 Yrs. kVp
Anterior Posterior	5	70	72	76	78
Lateral	10	70	78	82	88
Oblique	7.5	70	72	76	78

CHEST

CRONEX Hi-Plus Screens
Table Top
40 Inch (102 cm) Focal-Film Distance

	mAs	0 Mos. to 3 Mos. kVp	3 Mos. to 2 Yrs. kVp	2 Yrs. to 5 Yrs. kVp	5 Yrs. to 12 Yrs. kVp
Posterior	1.6	70	72	76	78
Lateral	3.3	70	78	82	88
Oblique	2.5	70	72	76	78

RIBS

CRONEX Hi-Plus Screens
12:1 Grid Ratio
40 Inch (102 cm) Focal-Film Distance

USE PEDIATRIC KILOVOLTAGE SCALES (PAGE 24)

	mAs
Anterior-Posterior	6.6
Oblique	10

Note: Measuring references would be like those for adults (see page 9).

BARIUM STUDIES

CRONEX® Hi-Plus Screens
12:1 Grid Ratio
40 Inch (102 cm) Focal-Film Distance

	kVp
Gastro-Intestinal	110
Barium Enema	110

	0 Mos. to 3 Mos. mAs	3 Mos. to 2 Yrs. mAs	2 Yrs. to 5 Yrs. mAs	5 Yrs. to 12 Yrs. mAs
Posterior Anterior	1.6	2.5	3.3	5
Oblique	2.5	3.3	5	6.6
Lateral	3.3	6.6	10	15

SKULL

CRONEX Hi-Plus Screens
12:1 Grid Ratio
40 Inch (102 cm) Focal-Film Distance

	mAs	0 Mos. to 3 mos. kVp	3 Mos. to 2 yrs. kVp	2 yrs. to 5 yrs. kVp	5 yrs. to 12 yrs. kVp
Towne	12.5	80	84	88	92
Posterior Anterior	12.5	74	78	82	86
Lateral	12.5	64	68	72	76
Basal	12.5	84	88	92	96
Waters	12.5	78	82	86	90

MANDIBLE

40 Inch (102 cm) Focal-Film Distance

Anterior (Grid Ratio: 12:1) (Hi-Plus Screens)

mAs	0 Mos. to 3 mos. kVp	3 Mos. to 2 yrs. kVp	2 yrs. to 5 yrs. kVp	5 yrs. to 12 yrs. kVp
12.5	70	74	78	82

Lateral (Table Top) (Detail Screens)

mAs	0 Mos. to 3 mos. kVp	3 Mos. to 2 yrs. kVp	2 yrs. to 5 yrs. kVp	5 yrs. to 12 yrs. kVp
12.5	58	62	66	70

NOSE

40 Inch (102 cm) Focal-Film Distance

Lateral (Table Top) (Detail Screens)

mAs	0 Mos. to 3 mos. kVp	3 Mos. to 2 yrs. kVp	2 yrs. to 5 yrs. kVp	5 yrs. to 12 yrs. kVp
3.3	58	62	64	68

Use Extension Cone

SINUSES

CRONEX Hi-Plus Screens
12:1 Grid Ratio
40 Inch (102 cm) Focal-Film Distance

	mAs	0 Mos. to 3 mos. kVp	3 Mos. to 2 yrs. kVp	2 yrs. to 5 yrs. kVp	5 yrs. to 12 yrs. kVp
Caldwells	12.5	78	82	86	90
Waters	12.5	78	82	86	90
Lateral	12.5	56	60	64	68

Use Extension Cone

MASTOIDS

CRONEX Hi-Plus Screens
12:1 Grid Ratio
40 Inch (102 cm) Focal-Film Distance

	mAs	0 Mos. to 3 mos. kVp	3 Mos. to 2 yrs. kVp	2 yrs. to 5 yrs. kVp	5 yrs. to 12 yrs. kVp
Towne	12.5	80	84	88	92
Schullers	12.5	70	74	78	82
Stenvers	12.5	66	70	74	78

Use Extension Cone

REFERENCE SOURCES RELATIVE TO REDUCTION OF ENTRANCE DOSE TO PATIENTS IN MEDICAL RADIOGRAPHY THROUGH THE USE OF FIXED AND HIGHER KILOVOLTAGE TECHNIQUES

1. **PHYSICAL FOUNDATIONS OF RADIOLOGY 2nd ED.,** 1954.
Glasser, Quimby, Taylor, Weatherwax. Medical Book Department of Harper & Brothers, 49 E. 33rd Street, New York, N.Y., pp. 171-172.

2. **LOW DOSAGE MEDICAL RADIOGRAPHY.**
Indiana State Board of Health, Indiana University Medical Center, Department of Radiology, Indianapolis, Indiana, 1964.

3. **MEASUREMENT OF BONE MARROW AND GONADAL DOSE FROM THE CHEST X-RAY EXAMINATION AS A FUNCTION OF FIELD SIZE, FIELD ALIGNMENT, TUBE KILOVOLTAGE AND ADDED FILTRATION.**
Epp, Weiss, Laughlin, Brit, J. Radiol. 34:85-100, February 1961.

4. **HIGH KILOVOLTAGE RADIOGRAPHY.**
Trout, Graves, Slauson, Radiology, 52:669-683, 1949.

5. **PRINCIPLES OF RADIOGRAPHIC EXPOSURE AND PROCESSING, 2nd ED.,** 1958.
Fuchs, A. W., Charles C. Thomas Publisher, Springfield, Illinois, p. 96.

6. **THE FUNDAMENTALS OF X-RAY AND RADIUM PHYSICS, 4th ED.,** 1965.
Selman, J., Charles C. Thomas Publisher, Springfield, Illinois, p. 425.

7. **2.4 GENERAL GUIDELINES IN THE CLINICAL USE OF RADIATION, NCRP Report No. 33**
Issued 2/1/68. NCRP Publications, P.O. Box 4867, Washington, D.C. 20008, p. 5.

mAs CONVERSION TABLES FOR QUANTA SCREENS

The introduction of QUANTA screens further advances low dosage radiography, permits stop motion technique, and can improve geometry through use of smaller focal spots at increased focal-film distance.

The speed of these receptors can be appreciated through direct comparison of milliampere-second values. To convert from Par Speed or Hi-Plus to rare earth, simply locate a known value, and directly read what would be used with the Quanta screen.

PAR SPEED	HI-PLUS	QUANTA II*	QUANTA III*
2.5	1.6	.8	.6
5	2.5	1.2	.8
6.6	3.3	1.6	1
8.3	4.2	2	1.2
10	5	2.5	1.6
13.3	6.6	3.3	2
16.6	8.3	4.2	2.5
20	10	5	3.3
25	12.5	6.6	4.2
30	15	7.5	5
40	20	10	6.6
50	25	13.3	8.3
60	30	15	10
80	40	20	13.3
100	50	25	16.6
120	60	30	20
150	80	40	25
200	100	50	33.3
250	120	60	40
300	150	80	50
400	200	100	70
500	250	120	80
600	300	150	100
800	400	200	120

*66-116 kVp

CRONEX® MEDICAL X-RAY FILMS AND BLUE-EMITTING INTENSIFYING SCREENS FOR RADIOGRAPHIC IMAGING

Regardless of your particular diagnostic procedure or your individual preference for speed or detail . . . Du Pont has the film/screen combination you need. Many of these combinations are based on Du Pont's newest imaging products — QUANTA III rare earth and QUANTA II blue-emitting intensifying screens, CRONEX 7 film and CRONEX 6 Plus film. These products were developed to provide new film/ screen options to meet today's specialized needs.

It's all part of Du Pont's commitment to enlarge your radiographic choices through ongoing research and development. Your Du Pont Technical Representative can discuss the wide array of choices now available.

CRONEX MEDICAL X-RAY FILMS	CRONEX BLUE-EMITTING INTENSIFYING SCREENS			
	NEW Quanta III	Quanta II	Hi-Plus	Par
Cronex 4	Speed **800** **Cronex 4/Quanta III** Ultra speed, image similar to Cronex 4/Quanta II with minimal increase in noise	Speed **400** **Cronex 4/Quanta II** High speed, image similar to Cronex 4/Hi-Plus	Speed **200** **Cronex 4/Hi-Plus** Industry standard at medium speed, high contrast.	Speed **100** **Cronex 4/Par** Industry standard at speed, high contrast, best sharpness.
Cronex 2DC	Speed **800** **Cronex 2DC/ Quanta III** Ultra speed, high contrast, image clarity of Cronex 2DC.	Speed **400** **Cronex 2DC/ Quanta II** High speed, high contrast, image clarity of Cronex 2DC.	Speed **200** **Cronex 2DC/Hi-Plus** Medium speed, high contrast. Industry standard for 3½ minute processing.	Speed **100** **Cronex 2DC/Par** Industry standard at Par speed and 3½ minute processing, best sharpness.
Cronex 6 Plus	Speed **800** **Cronex 6 Plus/ Quanta III** Ultra speed, excellent low density contrast plus tissue visibility.	Speed **400** **Cronex Plus/ Quanta II** High speed, excellent low density contrast plus tissue visibility.	Speed **200** **Cronex 6 Plus/ Hi-Plus** Medium speed, excellent low density contrast plus tissue visibility.	Speed **100** **Cronex 6 Plus/Par** Par speed, excellent low density contrast plus tissue visibility, best sharpness.
Cronex 6	Speed **800** **Cronex 6/Quanta III** Ultra speed, wide latitude, medium contrast.	Speed **400** **Cronex 6/Quanta II** High speed, wide latitude, medium contrast.	Speed **200** **Cronex 6/Hi-Plus** Medium speed, wide latitude, medium contrast.	Speed **100** **Cronex 6/Par** Par speed, wide latitude, medium contrast, best sharpness.
Cronex 7	Speed **400** **Cronex 7/Quanta III** High speed, less noise than Cronex 4/ Quanta II, image clarity like Cronex 2DC.	Speed **200** **Cronex 7/Quanta II** Medium speed, high contrast, image clarity like Cronex 2DC.	Speed **100** **Cronex 7/Hi-Plus** Par speed, high contrast, lowest noise, image clarity like Cronex 2DC.	

Use CRONEX Cassettes with 19 film/screen combinations from Du Pont.